Schriftenreihe des Lehrstuhls Kraftfahrzeugtechnik

Herausgeber Prof. Dr.-Ing. Günther Prokop

Band 10

ISSN 2509-694X

I0131720

Fakultät Verkehrswissenschaften „Friedrich List" – Institut für Automobiltechnik Dresden – IAD

Lehrstuhl Kraftfahrzeugtechnik

DOCTORAL THESIS

to attain the degree of Doktoringenieur (Dr.-Ing.)

PHYSICAL UNDERSTANDING OF
TIRE TRANSIENT HANDLING BEHAVIOR

Author:

Pavel Sarkisov
born on 10.07.1988 in Moscow

Evaluators:

Prof. Dr.-Ing. Günther Prokop
Lehrstuhl Kraftfahrzeugtechnik
Technische Universität Dresden

Univ.-Prof. Dr.-Ing. habil. Michael Kaliske
Institut für Statik und Dynamik der Tragwerke
Technische Universität Dresden

Submission date: 19.03.2018

Examination date: 18.04.2019

Dresden, 2018

Bibliografische Information der Deutschen Nationalbibliothek

Die Deutsche Nationalbibliothek verzeichnet diese Publikation in der
Deutschen Nationalbibliografie; detaillierte bibliographische Daten sind im Internet
über http://dnb.d-nb.de abrufbar.

1. Aufl. - Göttingen: Cuvillier, 2019
   Zugl.: (TU) Dresden, Univ., Diss., 2019

© CUVILLIER VERLAG, Göttingen 2019
   Nonnenstieg 8, 37075 Göttingen
   Telefon: 0551-54724-0
   Telefax: 0551-54724-21
   www.cuvillier.de

   ISBN 978-3-7369-7013-7
   eISBN 978-3-7369-6013-8

Acknowledgement

This thesis is a result of interdisciplinary research work that could not have been accomplished by the author alone.

Through scientific cooperation, discussions, and in experimental work, the author was supported, corrected and inspired by various people, whose contribution is highly appreciated.

First, I would like to express my gratitude to the supervisors of my research: Prof. Günther Prokop and Dr. Sergey Popov, who gave me valuable feedback. I am thankful to Prof. Michael Kaliske for his willingness to support as the second evaluator and to Dr. Denise Beitelschmidt for methodological assistance.

The possibility to work in two different universities assisted me greatly: On the one hand, I thank Dr. Sebastiaan van Putten, Hendrik Abel and Keun-Wook Chung for discussions in areas of vehicle and tire dynamics. On the other hand, my gratitude is addressed to Prof. Gennadiy Gladov, Prof. Georgiy Kotiev and Dr. Andrey Kupriyanov for the analysis in rolling theory and tribology.

All the developed experimental concepts were technically implemented, which could not have occurred without the great job done by Dirk Schlimper, Stefan Eckert, Erik Deunert and Axel Gerhard. I am also thankful to Jan Kubenz and Axel Wildgrube, who made it possible to produce samples.

As comprehensive research exceeds the area of competence of a single person, I would like to express my gratitude to people who have supported me by sharing their experience: Prof. Jörg Wensch, Prof. Peter Ruge, Mirko Riedel, Dr. Holger Rudolph, Rainer Barth and Tony Weber. These people have made it possible to work in an interdisciplinary and efficient manner.

My gratitude also goes to students who have contributed to the project by their diligence and creativity: Thomas Thüringer, Christian Betz, Philipp Ulbricht, Steffen Drossel and Tobias Mäbert.

I would like to mention the organizations who believed in potential of this research. The grants from Erasmus Mundus Action 2 MULTIC, Gesellschaft von Freunden und Förderern der TU Dresden, Automobil Forschung Dresden GmbH, TU Dresden Graduate Academy and DAAD are gratefully acknowledged.

In conclusion, I thank my friends and my family for supporting me.

Abstract

The development of advanced driving assistance systems and autonomous driving places challenging requirements on tire science: Transient rolling with combined slip and high slip values must be described. Because of increasing vehicle performance requirements and virtualization of its development process, it is required not only to know, which force is generated by tire, but also to know, how it happens.

Literature analysis reveals that no suitable tire model currently exists to fulfil these requirements: Empirical models provide no understanding, finite-element-models are too complex, simple physical models have, to date, been targeted towards different applications, but not towards understanding.

Further literature analysis has detected two relevant issues that require deeper investigation: Bending behavior of the tire carcass body and contact patch shape change in different rolling conditions, including the influence of these effects on tire transient behavior.

Hence, the goal of this research is to improve current understanding of the physical background of tire transient handling behavior, considering detected weaknesses.

The first step involves answering the question: What happens in a rolling tire?

Using acceleration measurement on the tire inner liner it was observed that the contact patch shape of the rolling tire changes nonlinearly with slip angle and becomes asymmetric. Optical measurement outside and inside the tire has clarified that carcass lateral bending features significant shear angle.

The second step answered the question: How can these processes be reproduced?

A simulation model with physically justified carcass and tread descriptions was developed considering the observed effects. A stable algorithm for numerical computing was designed. The model was qualitatively validated not only by output data (forces), but also by state parameters (deformation).

The third step clarified the issue: How do specific properties of tire influence its behavior and why?

The model-based analysis explained which tire structural parameters are responsible for which criteria of tire performance. It was found that both contact patch shape change and carcass shear angle have low influence on tire lateral force, but have perceptible impact on aligning torque generation.

Investigation of carcass lateral damping, consideration in the model of frequency-dependent material and friction properties of tread were identified as promising subjects for further research.

The main scientific contribution provided by this investigation is an improvement of understanding of tire physics in transient handling. This is required in the present day in order to support vehicle development process, to increase driver assistance systems efficiency and to improve road traffic safety.

Kurzfassung

Physikalisches Verständnis vom transienten fahrdynamischen Reifenverhalten

Die Entwicklung der Fahrerassistenzsysteme und des autonomen Fahrens bringt anspruchsvolle Herausforderungen an die Reifenwissenschaft mit sich. Eine von ihnen besteht darin, die Kräfte am Reifen beim transienten Rollen mit kombiniertem Schlupf und großen Schlupfbeträgen zu kennen. Wegen der steigenden Performance-Anforderungen und der Virtualisierung der Fahrzeugentwicklung ist es nötig, nicht nur erzeugte Reifenkräfte zu kennen, sondern auch deren Entstehung zu verstehen.

Die Fachliteratur bietet bisher kein geeignetes Modell für diese Anforderungen. So liefern empirische Modelle kein ausreichendes Verständnis der Prozesse, Finite-Elemente-Modelle sind zu komplex und einfache physikalische Modelle zielen eher auf die Anwendung anstatt auf das Verständnis der Prozesse im Reifen ab. Zudem wurden zwei relevante, aber noch nicht ausreichend untersuchte Gebiete der physikalischen Reifenmodellierung identifiziert: Das Biegeverhalten der Reifenkarkasse und die Form des Reifenlatsches in verschiedenen Fahrzuständen, inklusive des Einflusses dieser Effekte auf das transiente Reifenverhalten.

Infolgedessen ist das Ziel dieser Arbeit die Verbesserung des Verständnisses vom physikalischen Hintergrund des transienten fahrdynamischen Reifenverhaltens. Die Untersuchung erfolgt in drei aufeinander aufgebauten Abschnitten:

Der erste Abschnitt bezieht sich auf die Frage, welche Vorgänge im rollenden Reifen stattfinden.

Durch Messungen der Beschleunigung an der Innenfläche des Reifens wurde ermittelt, dass sich die Latschform des rollenden Reifens nichtlinear mit dem Schräglaufwinkel ändert. Optische Messungen außerhalb und innerhalb des Reifens zeigten einen signifikanten Schubwinkel des lateralen Biegeverhaltens der Reifenkarkasse.

Der zweite Abschnitt zielt auf die Frage hin, wie die Prozesse nachgebildet werden können.

Es wurde ein physikalisches Simulationsmodell unter Berücksichtigung der untersuchten Effekte entwickelt. Ein stabiles iteratives Berechnungsverfahren wurde erarbeitet. Das Modell wurde mittels der Ausgangsdaten (Kräfte) und mittels der Zustandsparameter (Verformung) qualitativ validiert.

Der dritte Abschnitt beantwortet die Fragen, wie eine bestimmte Eigenschaft des Reifens dessen Verhalten beeinflusst und wie diese Wirkkette gebildet wird.

Die modellbasierte Analyse klärte auf, welche strukturellen Parameter des Reifens verschiedene Kriterien des Reifenverhaltens bestimmen. Sowohl die Änderung der Latschform als auch der Schub im lateralen Biegeverhalten der Reifenkarkasse haben geringen Einfluss auf die Seitenkraft des Reifens, beeinflussen aber erkennbar die Rückstellmomentgeneration.

Als vielversprechende Aufgabenfelder weiterer Forschung stellten sich die Untersuchung der lateralen Dämpfung der Reifenkarkasse sowie die Erweiterung des Modells mit frequenzabhängigen Material- und Reibungseigenschaften des Profils heraus.

Der wichtigste wissenschaftliche Beitrag dieser Arbeit ist die Verbesserung des Prozessverständnisses der Reifenphysik im Hinblick auf die transienten fahrdynamischen Eigenschaften. Dieses Verständnis ist aktuell notwendig, um den Prozess der Fahrzeugentwicklung zu unterstützen, die Effizienz von Fahrerassistenzsystemen zu steigern und Fahrsicherheit zu verbessern.

Table of content

1 Introduction

1.1 Thesis structure

This thesis is thematically divided into five chapters. The logical chain is depicted in Figure 1.1.

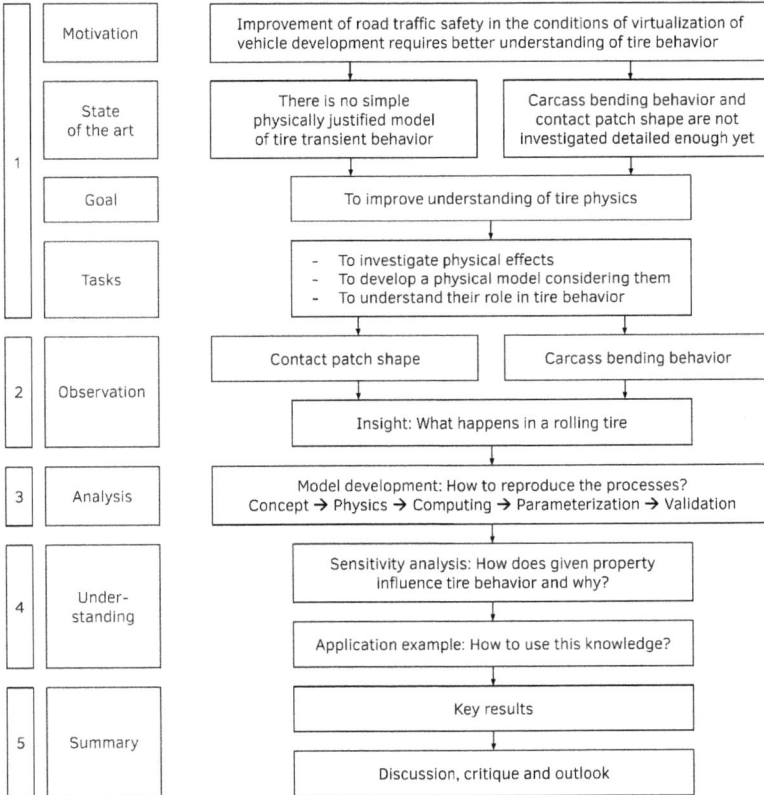

Figure 1.1. Logical chain of this thesis.

Three special symbols are used to highlight the following elements:

D   Definition: A statement of the meaning of a term used in this thesis.

?   Question: A step of logical argumentation, expressing the request of required information.

!   Insight: A clear understanding or conclusion regarding any relevant issue.

The core text of this dissertation is compressed down to the observations, arguments and conclusions necessary to achieve the aims of the study. Their background and further information are available in Appendix according to the references in the text.

1.2 Motivation

Road traffic injury is considered to be a global problem of healthcare and sustainable development: More than 1.2 million people die annually in road traffic accidents worldwide, and over 50 million are injured [WHO15]. Road traffic injuries are the main cause of death among people aged between 15 and 29 years, and are one of three leading causes among population aged between 5 and 44 years.

Due to the growing population and rising motorization rate, increasing attention is being paid to engineering solutions that can improve road traffic safety, like the seat belt and airbag did [DEK15]. Currently the highest potential for safety improvement is found in advanced driving assistance systems (ADAS, [Kno06]): These help a driver in critical situations through their sensing and acting abilities, which are not available for humans. Some of these systems have become legislatively mandatory, as occurred earlier with the seat belt.

Many driver assistance systems correct vehicle motion by adjusting wheel forces [Fen98]. The task of calculating the forces on four wheel hubs that are required to secure, for example, stability of vehicle motion, is relatively simple. However, the task to determine such an impact on the wheel that causes required force and torque response in a tire contact patch is considered to be a highly complicated problem of classical mechanics, which is not completely solved yet. Hence, improvement of road safety by means of ADAS is closely connected with tire behavior.

The ADAS, which correct vehicle motion by adjusting wheel forces in a critical situation, deal with a non-steady state. Steering angle may change stepwise during critical maneuvers. In addition to it, ADAS apply further dynamic interventions on the wheels, such as braking force (Figure 1.2).

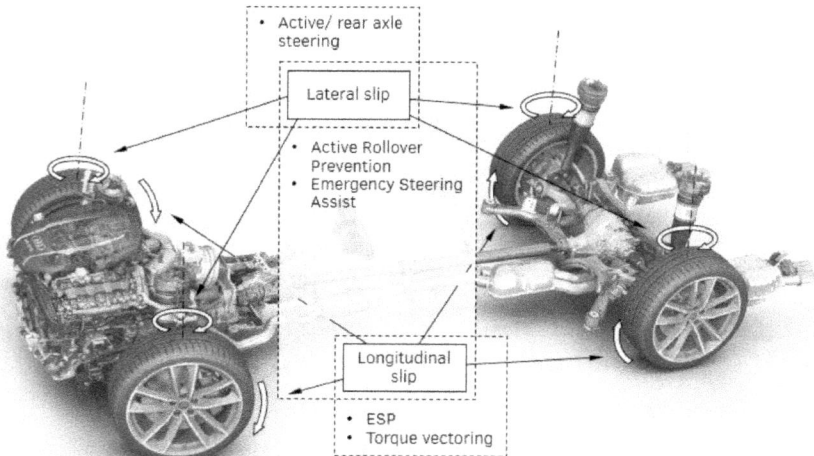

Figure 1.2. Impact of some assistance systems on wheels (background picture [AUD15]).

Electronic Stability Program (ESP, [Fen98]) impacts each wheel individually with a brake torque, modulated at up to 30 Hz, in order to stabilize the vehicle in critical situations such as the double lane change maneuver [ISO99, ISO11]. Taking into consideration that an electric vehicle with an individual drive is a realistic concept, in [Jal10, Jag15] it was proposed to substitute the individual brake torque intervention of ESP with individual drive torque for such vehicles. In this way, the maneuver can be finished faster, which means safer. Hence, both braking and driving torque can dynamically excite the tire.

Active steering systems [Lan15], which often include rear-axle steering [Har13] and active rear-axle kinematics [Wie08], combine high maneuverability at low speed with increased stability at high speed. These systems are also aimed to make vehicle reaction in critical situation predictable for driver [Don89, Köh06]. Different vehicle behavior in such situation compared to normal conditions is mainly caused by non-linear tire characteristic close to its friction limit. In [Yim15] it was proposed to adjust the slip angle of wheels individually in order to achieve the required yaw torque for vehicle stabilization with the lowest braking torque. Therefore, understanding of processes that are relevant for the non-linearity of tire characteristic (transient behavior [Hol99], rolling with combined slip and high slip values) is important for development of the active steering systems.

Active Rollover Prevention (ARP, [Sta14]) is a system that applies a full-power brake torque on a steered wheel and blocks it. In this manner the system avoids high lateral force as a cause of rollover.

Emergency Steering Assist [Har11] recognizes the possibility to avoid a crash by taking an evasive maneuver. If the driver initiates this maneuver, the system supports him/her with steering torque and individual brake interventions in order to ensure efficient evasion and stabilization of the vehicle.

Torque Vectoring systems [Saw96] manage the torque on each wheel to improve vehicle handling and its response in critical situation. The quality of vehicle state observation for optimal torque distribution is very sensitive to transient tire behavior and assumptions used for tire description in the observer [Bre95].

The possibility to use a **tire as a sensor** is emphasized by the latest developments in intelligent tire technology [Jo13]. With help of simple tire-mounted sensors (e.g., accelerometers) it is possible to estimate relevant data such as road conditions and tire wear level [Mas15]. The next goal is a real-time estimation of friction potential. Such data can significantly improve the efficiency of ADAS.

Finally, the concept of **autonomous driving** requires the vehicle to have an ability to act and react precisely and efficiently based on a number of real-time measurements. An efficient and stable control algorithm requires comprehensive understanding or at least description of a plant, which incorporates tire.

Because of increasing complexity of vehicles, early-decision-making becomes more important in vehicle development process (front loading). This leads to wider application of virtual tools and higher requirements on them. In the concept phase of development, simple and scalable simulation models are required, because chassis design data is known only at a very general level [Kva06, Wag17]. Understanding of physical processes in components and subsystems is essential for development of the corresponding simple model, aimed to an early stage of the development process [vPu17].

These facts can be summarized as observations, which raise specific requirements in the description of tire behavior (Table 1.2). The four considerations described in Table 1.2 formulate the scientific problem addressed in this research. Due to the increasing performance requirements to vehicles, virtualization of development process and rapid development of ADAS, this problem is relevant at the present day. The following section will therefore analyze the background of this problem and existing approaches to solving it.

Table 1.1. Detected challenges for tire description.

Observation	Requirements on tire description
ADAS adjust longitudinal force on the tire by means of drive or brake torque management; lateral force through steering intervention.	Consideration of combined longitudinal and lateral slip.
The highest efficiency of vehicle control corresponds to utilization of the entire friction potential of a tire.	Consideration of the entire range of slip values up to 100 %.
Electronic control systems operate at a frequency of up to 30 Hz during vehicle transient motion.	Description of transient behavior.
"Tire as a sensor" technology requires connection between simple measurements in the tire (e.g., acceleration sensor of Tire Pressure Monitoring System module) and relevant parameters such as road conditions or friction potential.	Understanding of physical processes.

1.3 State of the art

On June 10th, 1846, 23-years-old Robert William Thomson patented the "aerial wheel" [Joh17], which was the first mention of the pneumatic tire (Figure 1.3). Although he measured a 68 % rolling resistance reduction, nobody was able to develop this idea commercially at that time. In 1887, this concept appeared once again: A veterinary surgeon, John Boyd Dunlop, has fitted an inflated garden pipe to the wooden rim of his son's tricycle [Goo17]. Due to further application of this idea in cycling, pneumatic tire received recognition.

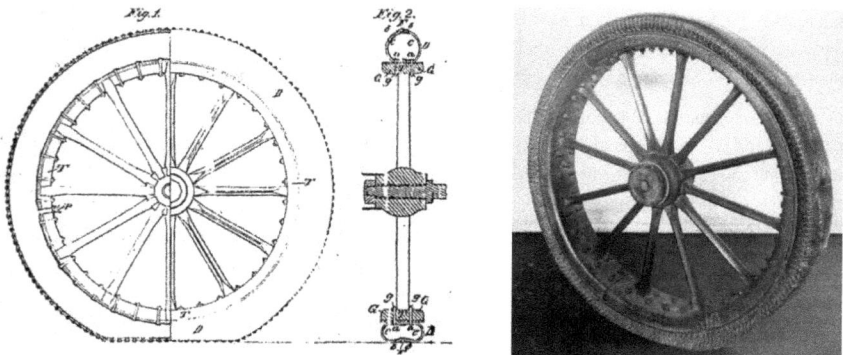

Figure 1.3. Patent US5104A by Robert William Thomson and his invention [Joh17].

The importance of the tire in the total vehicle system was recognized even during the very early periods of its development [Bro25]. Today, as one hundred years before, the tire remains the only link between a road surface and a vehicle. Still, despite its more than 150-year-long history, this physical system is not completely understood. Vehicle development process requires knowing a tire transfer function, which is usually called the "tire model".

In terms of handling, the "tire model" means a transfer function from kinematic state parameters to force parameters (Figure 1.4): Input parameters are motion laws describing wheel center coordinates and orientation angles in time. Output parameters, in terms of handling, are force laws describing longitudinal force, lateral force and aligning torque in time.

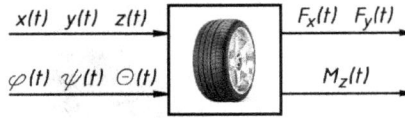

$$x(t) \quad y(t) \quad z(t) \qquad\qquad F_x(t) \quad F_y(t)$$
$$\varphi(t) \quad \psi(t) \quad \Theta(t) \qquad\qquad M_z(t)$$

Figure 1.4. The input-output scheme of the transient handling tire model.

Because of high complexity of tire behavior, the most common approach was to describe the tire as a black box, namely using empirical methods (Magic Formula [Bak91, Pac12]). Following the increasing scope of tire motion modes, which must be described, the empirical approach was enhanced with consideration of thermal issues [TNO13], inflation pressure [Sch05] and transient lateral force generation [Ein10]. However, it was also done empirically. In the meantime, present-day trends make empirical methods less suitable for modern requirements.

Firstly, continuously increasing performance requirements on modern vehicles lead to necessity to understand tire behavior further and to be able to use tire more efficiently. Secondly, vehicle development process experiences virtualization. An increasing share of simulation methods in vehicle development means that simulation models have to describe more complex physical systems and processes (in case of tire – combined slip, transient behavior). Comprehensive understanding is essential for this task.

Alternatively to the empirical characterization of tire handling properties, the physics behind them can be described with the help of FE-simulation [Bel00, Kal10]. Due to the highest level of detail, FE-modelling provides high accuracy, but is very time- and effort-consuming. It is reasonable to expend this effort for more complex applications, such as rolling over an uneven or deformable road surface [Cal15], vibration and comfort analysis [Bäc12], thermal processes [Bel97, Cal15b, Cal14]. Hence, for the handling analysis on rigid and flat road surfaces, an application of FE-modelling is less reasonable.

A predecessor of FE-analysis in tire dynamics is another important approach to understand the physical processes: To describe the tire with the help of a simple physical analog or mechanism. Apart from the FE-analysis, this option makes it possible to take into consideration only those physical properties and effects that are relevant to selected area of tire dynamics, and neglect the rest. This method provides an understanding of the processes and is suitable for complex motion, as long as a physical mechanism is able to respond to any correct excitation it receives. Based on these considerations, it is fair to select this approach to reach the goal of this study.

Traditionally for simple physical modelling, a tire is divided into two elements, the properties of which strongly differ: Tire tread layer (hereafter – 'tread') and tire carcass with the rest structural components (hereafter – 'carcass'). Properties, functions and constraints of these two elements are described in Table 1.3. 'Isotropy' indicates the uniformity of physical properties in different directions. If a deflection of one point of a flexible body causes deflection of other points, the body is considered to be cohesive. If a deflection of one point of the body does not cause deflection of other points (e.g. two tread blocks), such a body is considered to be non-cohesive.

Table 1.2. General separation of a tire into two structural elements.

	Complete tire	
	Carcass	Tread
Element	Tire carcass	Tire tread layer
Isotropy	Anisotropic: Composite structure	Isotropic: Homogeneous or close to homogeneous rubber
Cohesiveness	Cohesive	Non-cohesive
Function (in terms of handling)	• Longitudinal / lateral deflection • Lateral bending	• Longitudinal / lateral shear • Friction
External constraint	Flexible constraint with the rigid rim	Kinematic constraint with the road
Internal constraint	Fixed constraint with the tread	Fixed constraint with the carcass

Due to the different properties of these two elements, they are usually simulated with different simple physical structures. Because of cohesiveness and high lateral flexibility of the carcass, it is usually modeled as a body on an elastic foundation – rigid ring, flexible string or beam. The tread, on the contrary, is usually described via a brush model because of its non-cohesiveness. Following analysis emphasizes different directions of physical modelling with their advantages and disadvantages, in order to identify the most suitable method for the modern requirements.

The **string model** [Ell69] was initially considered to be a model for whole tire: The string in the free part of a tire was described using beam theory, the string in the contact patch was constrained kinematically. As a consequence, the model was unable to consider slip. An extension of this model [Böh66, Böh88] represents the tire as a beam on an elastic foundation and takes into account the mass of the carcass. Hence, this model considers inertia forces, beam bending and torsional stiffness. An analytic solution was deduced with help of Fourier transformation, but it is only available for the steady state.

The **brush model** ([Fro41, Pop93, Pac12]) describes a tire with radial bristles along its equator, which have a friction constraint with the road and can deflect in both horizontal directions (Figure 1.5). Due to isotropic brush properties, the model cannot properly describe the transient development of lateral force.

Although the string model and the brush model were initially developed to represent the whole tire, they are more suitable to describe carcass and tread respectively. This is why the most advanced physical tire models represent a combination of two structures. In [Böh85] a tire was considered as a wide string with a two-dimensional contact patch and brush elements on it. In such a manner, slip was taken into account. This model is able to consider transient behavior. However, its solution requires an expansion of functions into Taylor series, and there were used summands of only second order. This limits the accuracy and can be considered as a disadvantage of the method.

The **Brush and Ring Tire** [Gip97, Amm97] combines a rigid ring on an elastic foundation (carcass) with a brush model (tread). Based on the ring, a toroidal belt is constructed, which takes into account the two-dimensional contact patch (Figure 1.6). The elastic foundation is represented via non-linear spring-damper elements between a rigid rim and rigid ring. Brush elements of the tread model are massless, but have flexibility and damping. Still, the whole model is aimed at simulating the transient response of the tire to the vertical excitation (road unevenness). For the purpose of the simulation of a transient response to slip angle excitation, the carcass description as a rigid ring is a significant limitation: Its deflection is calculated quasi-static and deflected form is limited by a "rigid ring" assumption.

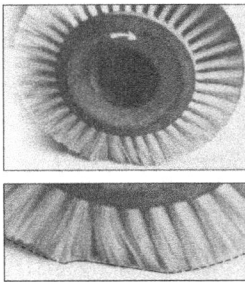

Figure 1.5. Structural scheme of brush model [Pop93, Pac12].

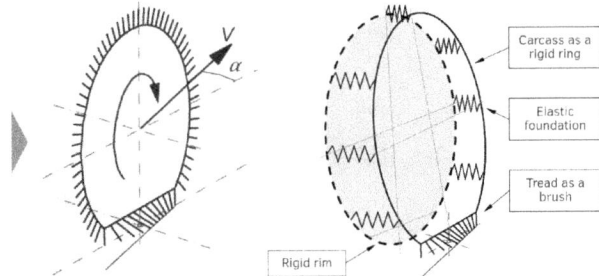

Figure 1.6. Structural scheme of the Brush and Ring Tire model in sense of lateral dynamics.

The **extended string model** [Hig97, Pac12] uses string on an elastic foundation to describe the carcass, and the brush model to simulate tread (Figure 1.7). Such a string-brush system can be multiplied in order to consider the width of tire. The entire circumference of the string is divided into two parts: The string in a contact patch and the string in a free part of the tire (Figures 1.8-1.9). These two parts are connected once on the leading (A) and once on the trailing (B) edges of the contact patch.

The model has two important assumptions. Firstly, the connection between the strings at point B is not limited by tangency. Consequently, the carcass has a kink in this point.

Secondly, the orientation of the string at point A is defined by an assumption that was described by [Böh66b]: In the middle plane of the tire there is constructed point D, which is distanced from leading edge (point L) by a relaxation length. At point D, a pivot is assumed. Next, an imaginary lever with a roller is considered, which pivots around point D. The roller always remains on the leading edge of the contact patch. This non-holonomic constraint defines the orientation of the string at point A.

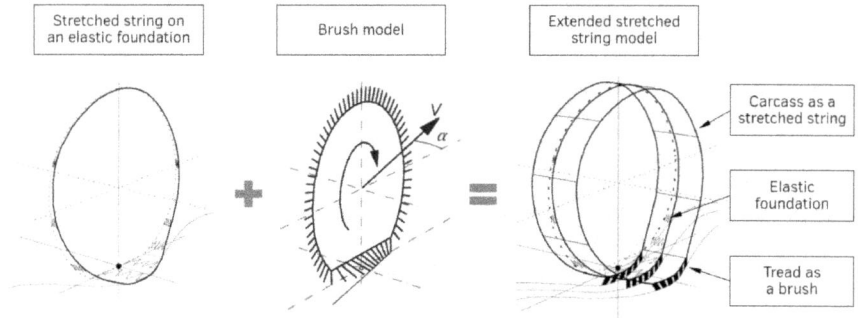

Figure 1.7. Structural scheme of the extended string model [Pac12].

These assumptions make model calculation fast. However, they greatly influence the description of the carcass, which is important for transient aligning torque generation. In addition to this, the representation of the wide carcass via several independent one-dimensional strings limits proper reproduction of carcass bending behavior.

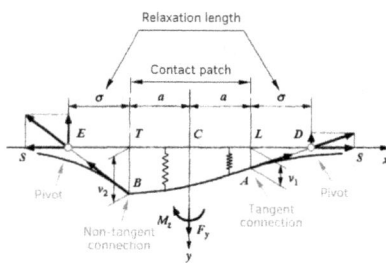

Figure 1.8. Top view on the string model, a representation of [Hig97].

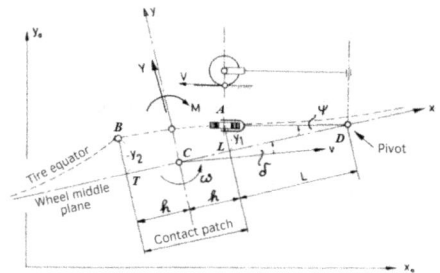

Figure 1.9. Top view on the string model, a representation of [Böh66b].

The **Treadsim** model [Pac12, Uil07, Hoo05, Fia54] considers carcass deflection only in the contact patch. Initially, this deflection was approximated by a second-order function [Pac12] (Figures 1.10-1.11). Later, it was changed to beam elements [Hoo05] (Figure 1.12). The tread is described with the brush model.

Even in its first representation, the Treadsim model had already achieved a level of complexity that made it impossible to obtain an analytical solution and required a numerical solving routine. This lead to oscillatory stability issues inherent in iterative solving process, which lead to divergence [Pac71].

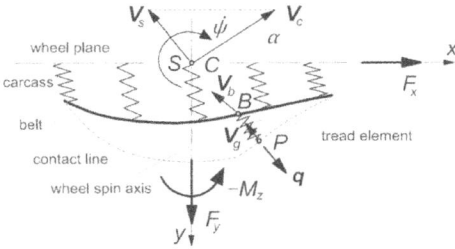

Figure 1.10. Structural scheme of the Treadsim model [Pac12].

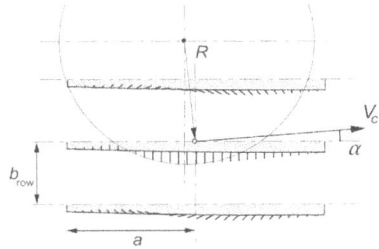

Figure 1.11. Multirow representation of the Treadsim model [Pac12].

In order to consider the width of the tire and be able to describe transient aligning torque generation, the model multiplies the beams with the brush along the tire axis (Figure 1.11). Still, these beams remain independent from each other; this fact limits precise representation of the bending behavior of the wide carcass, which is especially important for aligning torque generation. Even in a steady state, simulation error of aligning torque can be as high as 25 % [Uil07].

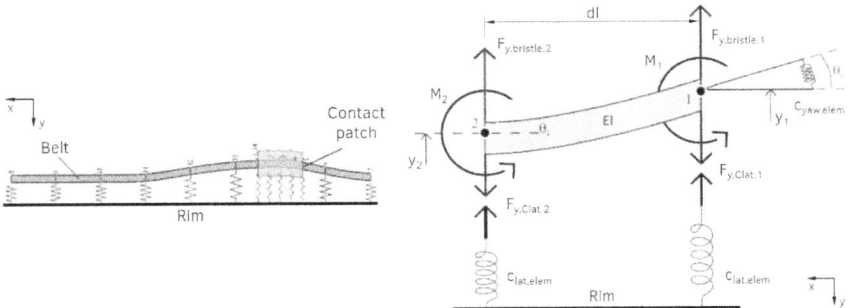

Figure 1.12. An extension of the Treadsim model with a beam-based carcass [Hoo05, Uil07].

The **TameTire** model [Fév07, Fév08, Fév10] is one of the most advanced physical models, combining mechanical description with a thermal model. The tread is described once again as a brush model. Carcass deflection is simulated with help of a second-order approximation function, based on lateral force, tire lateral flexibility and kinematic constrains (Figure 1.13).

It is noteworthy that this function has the same appearance as that of the Treadsim model. This fact confirms that the second-order function is a good compromise between simulation accuracy and computational effort. Consequently, the model is real-time capable. The price of this advantage is the simplification of the carcass deflection form and an empirical description of transient behavior.

The second-order approximation function limits carcass simulation in cases of high-frequency excitation, especially for aligning torque. The empirical description of transient behavior leads to the black-box based simulation of transient processes, which does not provide an understanding of the physical background.

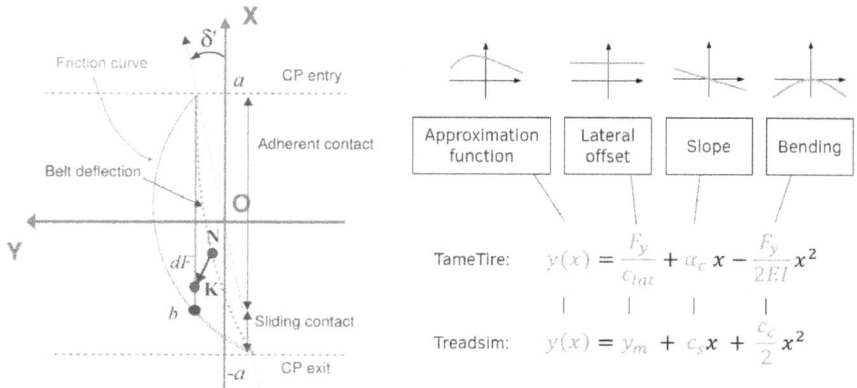

Figure 1.13. Structural scheme of the TameTire model and the composition of the carcass deflection approximation function [Fév08].

A comparison of these four described physical models emphasizes that each is aimed at different target application (Table 1.4). As long as the task of this research, which is determined by new trends in the automotive industry, differs from the target applications of the analyzed models, it is natural that they are unable to meet new requirements. However, there is a lot to be learned from them.

The goal of this investigation is to understand the physical background of the transient handling properties. In order to be able to identify a connection between the physical effects of the rolling tire and its transient handling behavior, it is important to describe the most relevant effects physically (not empirically) and physically correct (without qualitative contradiction such as a carcass kink). This is why both carcass and tread models have to be physically justified. The simplification of the wide carcass structure affects the accuracy of the aligning torque description [Uil07]. Hence, it is necessary to consider the wide common body of the carcass, analyze its bending behavior and take into account the two-dimensional array of brush elements.

A commonality in the described physical models is that each uses several important physical properties of the rolling tire to describe its behavior:

- Lateral flexibility of tire carcass, usually considered via elastic connection between tire carcass and rigid rim;
- Carcass bending, usually taken into account via flexible one-dimensional elements such as string or beam, or just approximated with a given assumed function;
- Tread shear, usually simulated via brush model;
- Friction properties between tread and road.

Due to importance of these properties for tire handling, they are considered in this study as **primary physical properties**. They make it possible to describe the transient generation of lateral force and aligning torque (partly). However, they do not provide the required understanding of the physical background, because of several simplifications.

Table 1.3. A comparison of physical tire models in sense of requirements of the current research.

	Target application	Carcass model	Tread model	Out-of-plane extension	Incompatibility with requirements
Requirements	Understanding of physical background of transient handling behavior	Physically justified deflection: No geometric limits to form, no kinks	Physically justified shear and friction model	Wide carcass body considering its bending behavior	-
Brush and Ring Tire	Dynamic force and torque response to short-wave road unevenness	Rigid ring on an elastic foundation	2D brush model	Extension via toroidal ring shell	- Constrained carcass deflection (rigid ring); - Quasi-static change of carcass deflection
Extended string	Simulation of transient handling properties at large slip and camber; vibration analysis	Stretched string on an elastic foundation	1D brush model	Multiplication of in-plane models	- Constrained carcass deflection (kink, orientation); - Simplified out-of-plane extension (several independent strings)
Treadsim	Analysis of the influence of considered physical effects on transient handling properties	Beam on an elastic foundation	1D brush model	Multiplication of in-plane models	- Simplified out-of-plane extension (several independent beams)
TameTire	Real-time simulation of transient handling properties	Second-order approximation function with an elastic foundation	1D brush model	Wide carcass body; No data about bending description	- Simplified carcass deflection (second-order curve)

In order to obtain deeper insight into the processes in the rolling tire and to better understand force and torque generation, it is necessary to take into account further physical effects and properties, as done in [vPu12], for example: There was investigated the influence of a two-dimensional contact patch on transient aligning torque generation.

An analysis of tire structure, rolling processes and the latest developments in tire modelling has identified two areas of scientific knowledge that are of importance in understanding of tire handling properties, but have not yet been sufficiently investigated.

The first reasonable area for research is justified by the fact that a contact patch is the only area on which the tire-road interaction forces and torques are transmitted. As the shear deformation of the tread blocks develops in a nonuniform way from the leading to the trailing edges, the form of these edges influences lateral force and aligning torque. Consequently, the contact patch shape is not irrelevant for tire handling properties.

The FEM-simulation [Gim01, Gim07, Jo13, Cal15] and the dynamic contact patch measurement [Kuw13, Kuw14] have shown that even small values of slip angle (1°) cause significant change in the contact patch shape: It becomes asymmetric (Figure 1.14).

However, the influence of this change on tire transient handling properties has not yet been considered. Therefore, it is reasonable to research this event chain.

Justification of the second area deserving further research is based on the following fact: As shear stiffness of tire tread is approximately one order of magnitude higher than carcass lateral stiffness, the carcass lateral deflection is extremely important for force and torque generation. Previously a flexible carcass was described with one or several independent one-dimensional strings or beams on an elastic foundation (Table 1.4). The deformation behavior of such a system is different to the deformation of wide common carcass body.

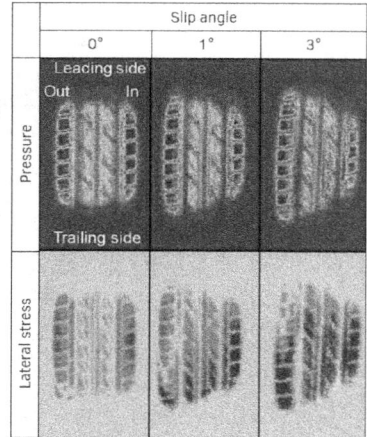

Figure 1.14. Strain figure in contact patch of a cornering tire [Kuw14].

Various researchers [Erd09, Erd11, Han13] have investigated the lateral deflection of the carcass equator or its centerline in order to understand its bending properties and their relationship with tire handling behavior. However, these works were focused only on the tire equator, assuming that the wide carcass body follows the equator. That raises the question:

> ? How does the wide body of a carcass actually deform: not only the equator, but also the fibers outside the equator?

The influence of this behavior on the transient handling properties of a tire has not yet been analyzed detailed enough too.

These issues are considered in this research as **secondary physical properties**. As they have not yet been investigated in sufficient detail in terms of transient handling behavior of a tire, they complete the mission statement of this research.

1.4 Mission statement

Based on set requirements, with the assistance of the experience brought by the existing models and using results of the literature analysis, it was possible to set a goal, to pose research questions, to select a method, to assign tasks and to limit a research area.

The **goal** of this research is to improve the understanding of the physical background of tire transient behavior. The particular mission means answering following **research questions**:

- Which physical effects and properties determine tire transient handling behavior?
- How can these effects and properties be reproduced in a simulation model?
- How does every single effect influence tire transient handling behavior? What is its role?

The reasonable **method** for this is decomposition of a rolling tire into separated physical effects and properties and their physical reproduction in a simulation model.

The **tasks** involved in achieving the goal can be divided into three categories:

1. **To observe:**

    1.1. To develop necessary experimental methods and to investigate in detail the secondary physical properties of a tire: contact patch shape change and carcass bending behavior;

    1.2. To measure the primary physical properties of the same tire: carcass lateral stiffness, tread shear stiffness, friction properties.

2. **To analyze:**

    2.1. To develop a physical simulation model that meets the following requirements:

        2.1.1. Physically justified carcass description, considering carcass as a wide common body;

        2.1.2. Physically justified tread description, including variable contact patch shape, but with a limited consideration of material and friction properties;

    2.2. To parameterize and to validate the developed model.

3. **To understand:**

    3.1. To investigate with the assistance of developed simulation tool the influence of different physical properties and effects of tire on its transient handling behavior.

Because of complexity of the tire rolling theory, it is fair to limit the area in this thesis by rolling on a flat rigid surface with transient slip angle or wheel load excitation without camber variation.

1.5 Main terms and hypotheses

The majority of the terms used in the thesis are common. However, some differences in interpretation may occur between European, American, Russian and Japanese automotive schools of thought. For clarity, the meaning of several key terms used in this thesis is specified.

Wheel with a pneumatic tire (hereafter – 'wheel') is a mechanical system, consisting of a rigid body of revolution (**rim**) and flexible shell (**tire**), which together circumscribe a certain closed subspace (inflated air).

Stationary wheel (tire) is a wheel (tire) which is not rolling, but can be excited in other directions.

Road coordinate system $O_R X_R Y_R Z_R$ is a Cartesian coordinate system fixed with the road, where longitudinal and lateral axes belong to the road plane (Figure 1.15). Origin O_R is any given point on the road surface.

Wheel coordinate system $O X_W Y_W Z_W$ is a Cartesian coordinate system with an origin in the rim center. The longitudinal axis $O X_W$ is parallel to the flat road surface. The lateral axis $O Y_W$ coincides with the wheel rotation axis.

Geometrical contact patch center C is an intersection of the $O Z_W$ axis with the road surface.

Contact coordinate system $C X_C Y_C Z_C$ is a Cartesian coordinate system with an origin in the geometrical contact patch center C, a longitudinal axis parallel to the wheel longitudinal axis $O X_W$ and a vertical axis oriented normally to the road plane. The forces and torques in the contact patch are considered in the sense of the contact coordinate system.

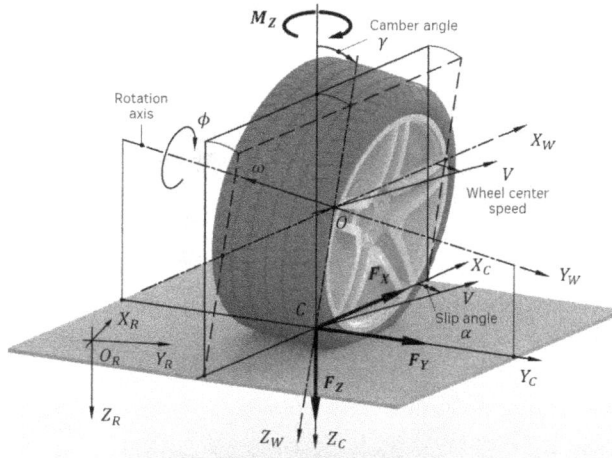

Figure 1.15. Adopted coordinate systems.

Parameters of rolling are considered to be six kinematic laws for six degrees of freedom, which uniquely determine the position of the rim at each moment in time: three translational coordinates of the rim center $x_O(t)$, $y_O(t)$, $z_O(t)$ and three orientation angles $\theta(t)$, $\psi(t)$, $\varphi(t)$. These parameters can be recalculated to slip angle $\alpha(t)$, camber angle $\gamma(t)$, rotation angle $\phi(t)$. Rather than $\phi(t)$, its derivative is commonly in use. It is rotation speed (angular velocity) $\omega(t) = \dot{\phi}(t)$.

Slip angle $\alpha(t)$ is an angle between the longitudinal axis of the wheel OX_W and the vector of the horizontal speed of the wheel center C.

Static rolling (steady-state rolling) is considered to be a rolling mode, where:

- Horizontal coordinates change with constant speed: $\ddot{x}_O = 0, \ddot{y}_O = 0$
- The vertical coordinate does not change: $\dot{z}_O = 0$
- Slip angle and camber angle do not change: $\dot{\alpha} = 0, \dot{\gamma} = 0$
- Wheel rotation angle changes with constant speed: $\ddot{\omega} = 0$

Quasi-static rolling is considered to be a rolling mode in which variations of rolling parameters exceed the conditions of static rolling, but they happen slowly enough that the system remains in internal equilibrium or close to it.

Non-steady-state rolling (dynamic rolling, transient process) is considered to be a rolling mode in which variations of rolling parameters exceed the conditions of static rolling and happen sufficiently quickly that the system moves out of equilibrium. The system requires time to gain deformation and to come close to the equilibrium. Depending upon this delay and its relevance for the given task, non-steady-state rolling must either be taken into account or be simplified to a quasi-static state.

All further considerations are formulated with terms with described meanings, unless otherwise specified.

1.6 Summary of chapter 1

The rapid development of ADAS, the coming soon technologies of autonomous driving and intelligent tire are enforcing new requirements on tire science. Transient rolling with combined slip and high slip values must be described. The connection between the processes in the tire and the data, which is relevant for vehicle dynamics, must be captured. Vehicle development process requires simple and scalable models for complex rolling conditions. Hence, to solve these problems, an understanding of physical processes in the rolling tire is required.

Empirical methods are unsuitable for this, because of the "black box" description. FE-methods are less applicable because of their high complexity. A reasonable tool to use is therefore a simple physical model.

Existing physical models are insufficiently compatible with the set requirements, because they were targeted differently, usually application-oriented rather than understanding-oriented. Still they provide a large amount of experience. An analysis of these models has detected several issues that can be changed in order to develop a suitable model for the set task (wide carcass body considering its bending behavior).

The literature analysis identified two areas of scientific knowledge that represent importance for understanding of tire handling properties, but which to date have been insufficiently investigated: Contact patch shape change while cornering and bending behavior of wide carcass. An influence of these effects on tire transient handling properties also was not researched detailed enough.

Now it is clear why to do (motivation), what to do (goal) and how to do (tasks and method). The next step is to gain the necessary knowledge through experimental observation.

2 Experimental investigation of tire deformation

2.1 Introduction to experimental research

Experimental analysis of the questions that were defined in the previous chapter requires the following inputs: test samples, testing equipment, testing methods and data analysis methods. The outputs of such an analysis are: general observations, qualitative and quantitative findings regarding the raised questions. Items from the both inputs and outputs are sequentially described in this chapter. The conclusion of it summarizes the most important insights for the next step, which is model development.

2.2 Test samples

The test object was a passenger car tire 255/35 ZR19 (summer configuration). It represented the typical low profile tire without any structural features such as Run-On-Flat sidewalls.

In order to be able to measure the properties of tire carcass and tread separately, the complete tire was divided into two essential elements (Figure 2.1): The tire carcass and tread layer.

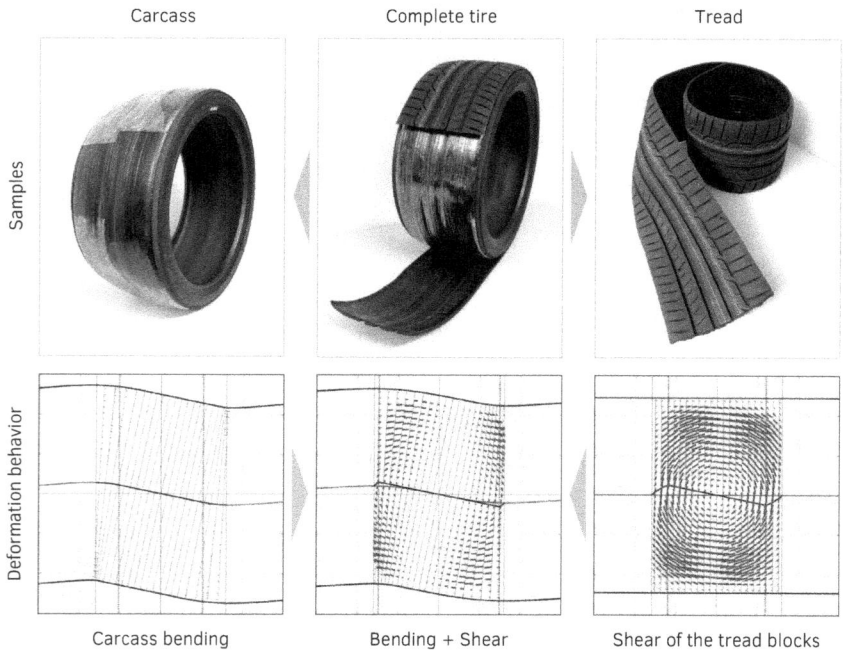

Figure 2.1. Deformation of the complete tire as a composition of deformation components of the carcass and tread (deformation is shown as a top view on the contact patch).

This division was performed without any damage to these elements. Before winding up the tread layer, the carcass was covered with a heat-resistant film. Next, the tread was wound up on this protected carcass. This workpiece was treated via vulcanization in the usual manner to achieve the same properties as an ordinary tire. As the film prevented diffusion between the tread and carcass outer layers, the tread belt was easily separated from the carcass (Figure 2.1).

The European Tyre and Rim Technical Organisation (ETRTO) [ETR06] specifies the recommended inflation pressure for this tire as 2.9 bar and a rim width of 9J, which were used in all tests. In order to prepare the tire for the tests, it was preconditioned according to the typical procedure: 45 minutes of rolling at 100 km/h with those values of inflation pressure, wheel load, and camber, which will be used in the measurements: respectively 2.9 bar, 5 kN, 0° [Gie12, p. 77].

2.3 Experimental equipment

The main experimental tool was the drum tire test rig of the Institute of Automotive Technologies Dresden at the Technische Universität Dresden (Figure 2.2, [Sar15]). The wheel was mounted on a horizontal frame that served to vary the camber angle. This assembly pivoted on a vertical frame and was connected to it via two hydraulic cylinders. The vertical frame was fixed with a major hydraulic cylinder. This frame enables rotation around vertical axis and vertical displacement of the wheel. The described assembly is mounted on a framework of a rig, which consist of four columns and a traverse.

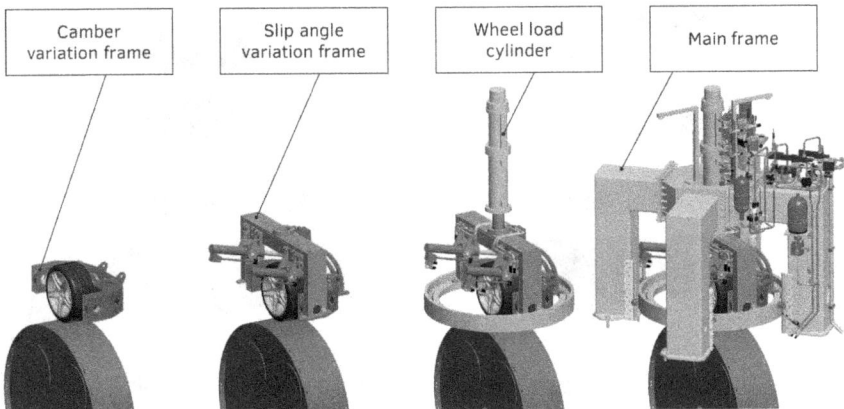

Figure 2.2. Drum tire test rig.

An important advantage of this test rig is the high rates of camber and slip angle variation (Table 2.1). This allowed to excite tire dynamically and to investigate its transient response.

A general problem of the drum tire test rig is the influence of a drum curvature. The primary background of this influence is difference in the contact patch length on flat and cylindrical surfaces [Unr13]. A comparison of tire footprints on a flat surface and on the given drum showed the difference of the contact patch length up to 20 % (Figure 2.3).

The correlation between on-drum measurements and on-road characteristics was not a focus of this investigation. Neither was the correlation between the simulation model and on-road characteristics, because the model was parameterized with on-drum measurements. The focus was understanding of the physical processes of a rolling tire regarding its handling behavior. As long as they are generally similar on a drum and on a flat surface, the described test rig is suitable for the research task.

To ensure the quality of the measurements, the test rig static stiffness was investigated [Sar15]. The stiffness accounted for 35 kN/mm in both longitudinal and vertical direction of a tire and 64 kN/mm in lateral. For an average tire, the error caused by test rig flexibility accounts for 0.8, 1.4, and 0.5 % in the vertical, longitudinal, and lateral directions, respectively.

Table 2.1. Drum tire test rig technical data.

Longitudinal speed	3 – 300 km/h
Wheel load	0 – 30 kN
Horizontal forces	−20...+20 kN
Slip angle	−90...+90°
Slip angle rate	50 °/s
Camber	−45...+10°
Camber rate	30 °/s
Drum diameter	2 m
Drum width	0.5 m

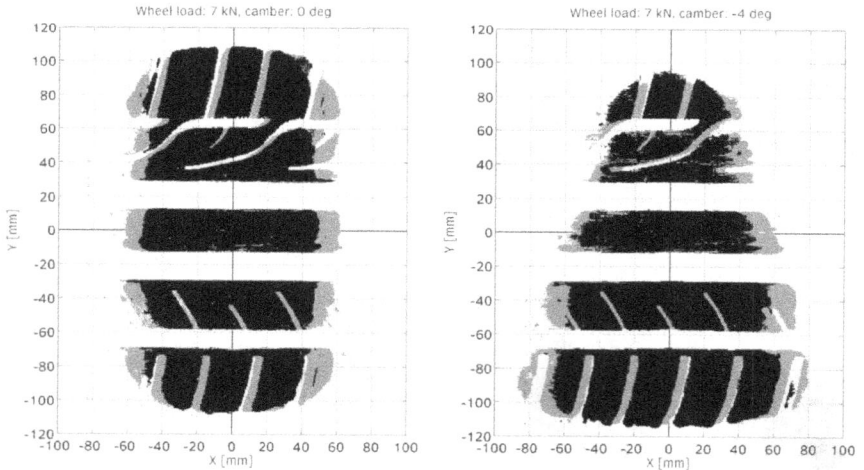

Figure 2.3. Footprints on the flat plate (gray) and on the 2-m drum (black).

Due to the ability of the test rig to dynamically vary tire slip angle, its high static stiffness and the low influence of the drum curvature on the general physical processes (not tire performance indicators), the described drum tire test rig is suitable for the investigation of the physical background of tire transient handling behavior.

For the analysis of material properties, a hydraulic pulsing machine was used, which is described in Appendix A.1.

The above-mentioned test rigs were the basic tools for the experimental work. In the next sections, experimental analyses of relevant physical properties of the tire will be introduced, including a brief description of the additional equipment used for these investigations.

2.4 Contact patch pressure distribution

The first analysis is focused to the limit of friction forces in the contact patch, which greatly depends upon pressure. The measurement was performed with pressure measuring film Fujifilm Prescale 4LW (precision of 10 %, Figure 2.4).

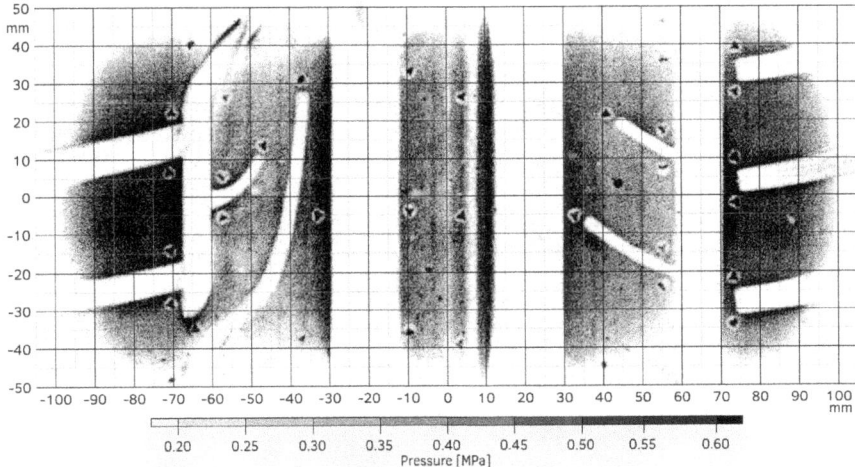

Figure 2.4. Pressure measurement in tire footprint on flat surface with the wheel load of 5 kN.

In the majority of the contact area, the contact pressure varied from 0.35 to 0.55 MPa. Pressure distribution in the longitudinal cross-sections of tire was close to uniform. Zones located along the edges of the contact patch featured a gradual reduction of pressure. A comparison of the pressure distribution on the drum and flat surface was performed on a qualitative level and showed a similar figure, but different length of the contact patch. Hence, simplified footprint measurements (which provide only contact area but no pressure information) are still suitable for the research tasks. The pressure distribution has to be assumed in this case.

According to [Sak95, Ahl12], it is acceptable to assume an uniform pressure distribution of a rolling tire in physical models, considering a gradual drop to zero in the vicinity of leading and trailing edges.

For creating such pressure distribution field, contact patch geometry is required. This geometry is the focus of the next subchapter.

2.5 Contact patch geometry of the rolling tire

2.5.1 State of the art

As shown in Figure 1.14, a shape of a contact patch while cornering can be estimated using finite element methods. This way, however, requires effort-consuming model parameterization and calculation. Another approach to estimate the contact patch shape is based on the measurement of its length with help of acceleration sensors. This method was initially used to investigate aquaplaning [Mat15, Nis14, Nis15] or to estimate rolling conditions, e.g. slip angle [Mas15, Tib13]. In order to answer the question, how the change of the contact patch shape influences lateral force and aligning torque, it is necessary to analyze not only contact patch shape, but also force and torque generation of the tire within the same run. A suitable tool for force and torque measurement is the drum tire test rig (Chapter 2.3). Analysis of the contact patch shape requires a special method.

2.5.2 Measurement method

According to [Mat15, Nis14, Nis15], three acceleration sensors located on half of the tire width provide an acceptable estimation of contact patch shape, but only if it is symmetric about the longitudinal axis. As the contact patch of a cornering tire is not symmetric, the number of sensors must be at least doubled. Considering the reference sensor in the middle plane of the tire, seven sensors are required. In order to ensure the definition of the contact patch length based on radial acceleration, the reference sensor must also measure longitudinal and lateral acceleration.

For this application, a wide-bandwidth acceleration sensors Analog Devices ADXL001-500BEZ (measuring range: ±500 g; nonlinearity: 2 %) were selected (Figure 2.5).

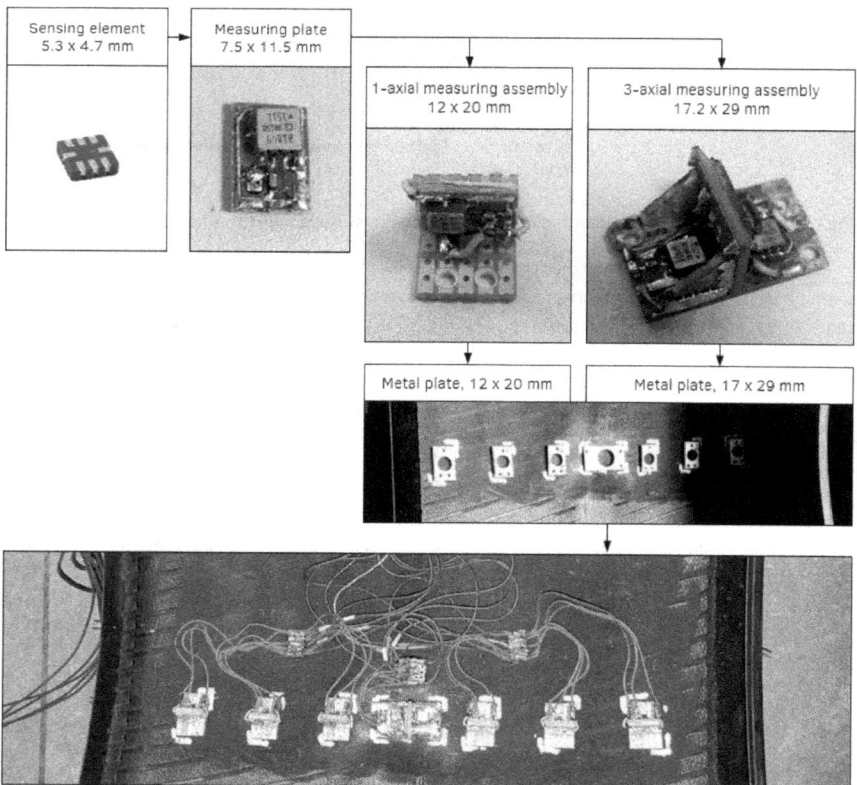

Figure 2.5. Sensors set up in the tire.

The bandwidth of 22 kHz made it possible to achieve a sampling rate as one measurement per 1.7 mm of tire travel in the worst case (120 km/h).

The sensing element was mounted on a measuring plate, which was used in two layouts: One plate on the frame (single-axial) and three plates on the frame (tri-axial, Figure 2.5). The frames of both types were mounted on metal plates with screws. The metal plates were glued to the tire inner surface. Overall, there were six single-axial measuring assemblies and one tri-axial measuring assembly.

During every revolution of the tire, the radial acceleration signal depicted specific reproducible form: Figure 2.6 shows sample measurement result for a complete tire (carcass with tread) and result for the carcass only; both rotated with 100 km/h and were loaded with 7 kN.

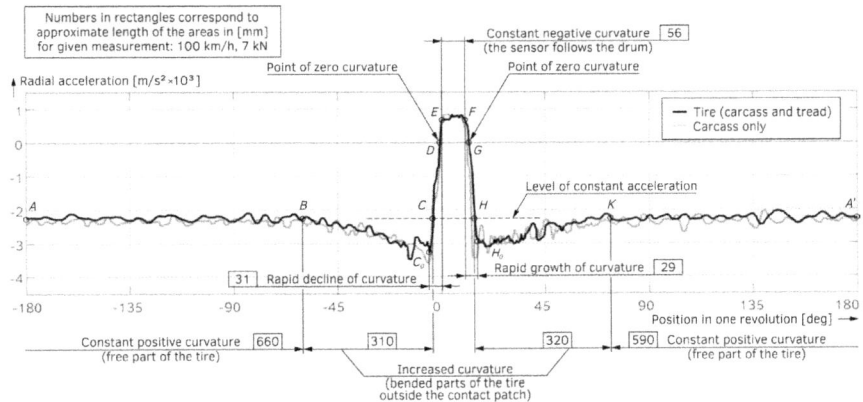

Figure 2.6. Measurement of radial acceleration in tire and carcass, one revolution.

As shown in Figure 2.6, the measurement signal had a form of a constant line with two steps. Due to this form, the frequency based filters were unsuitable. Hence, the median filter using a window size of 11 samples with one entry was selected.

In Appendix A.2, details concerning the origin of this signal are described. It was found that the used acceleration sensor assembly smoothed the rapid acceleration drop and rise because of its length. Hence, measurement with the applied acceleration sensors was not able to deliver reliable numerical value of the contact patch length. But it could deliver a range, in which the start point is located (CD) and a range, in which the end point is found (GH). These data are sufficient to follow the change of the contact patch form. In the following diagrams, there will be depicted positions of points C, D, G and H in the form of histograms, as it is the most useful information which is trustworthy, not assumed. Due to the symmetry of the drum, tire and sensor plate, with the highest possibility the start point is in the middle of CD; the end point is in the middle of GH. These assumed positions will hereafter be called 'conditional' (Figure 2.7):

D **Conditional start (end) of the contact patch** – a middle point of radial acceleration drop CD (rise GH), assumed to be a start (end) of the contact patch.

D **Conditional length of the contact patch** – the distance between the conditional start and end points.

The seven acceleration sensors will be indicated with capital letters A to G. For comparison, there is shown the footprint of the non-rolling tire on the same drum with the same wheel load.

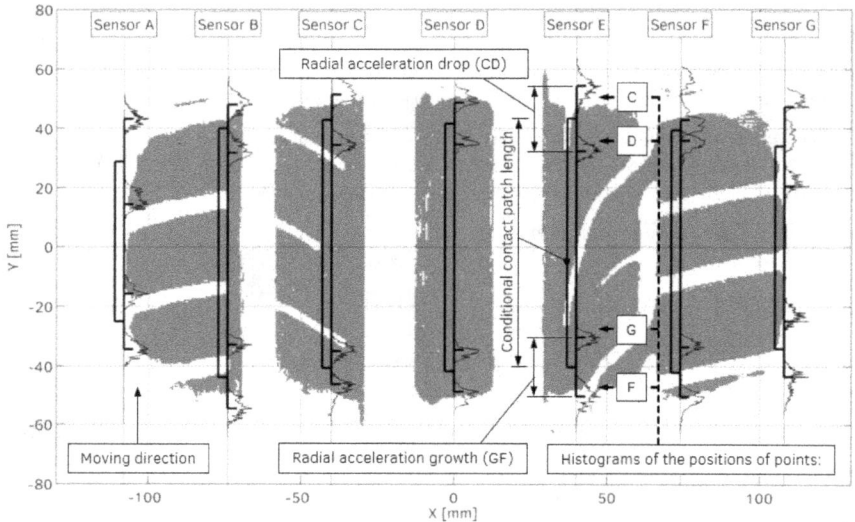

Figure 2.7. Measurement of contact patch length with acceleration sensors (wheel load is 5 kN).

Both samples (complete tire and tire carcass without tread) were analyzed using the described approach in order to trace the influence of the tread layer.

The difference between the static footprint and the contact patch length measured with acceleration sensors is not considered intentionally, because the values of the conditional contact patch length were sufficiently reliable to describe only the qualitative change of the contact patch shape. They were not reliable enough to detect the absolute value of the contact patch length.

The following subchapters summarize the observations of the contact patch shape in relation to different influencing factors. During each of the following tests, the vertical position of the wheel was continuously adjusted in order to maintain the wheel load on the same given level.

2.5.3 Contact patch geometry depending upon rolling speed

As a tire is a flexible body, its rotation causes radial deformation due to centrifugal forces. This effect changes not only the contact patch, but also the vertical reaction force. In order to maintain the wheel load on the same level for all values of rolling speed, the vertical position of the wheel was dynamically adjusted. Figure 2.8 shows that with increasing speed, the wheel had to be lifted for the purpose of maintaining the constant wheel load: The lift displacement per 100 km/h accounted for 1 mm for carcass and 1.5 mm for tire. This difference illustrates the influence of the tread mass.

Figure 2.9 shows the change of the conditional start and end of the contact patch in the middle of the tire (sensor D) depending upon the rolling speed. Although the vertical position of the wheel was adjusted and the wheel load was kept the same, both tire and carcass showed continuous growth of the contact patch length with the speed, at the rate of ca. 10-15 % per 100 km/h.

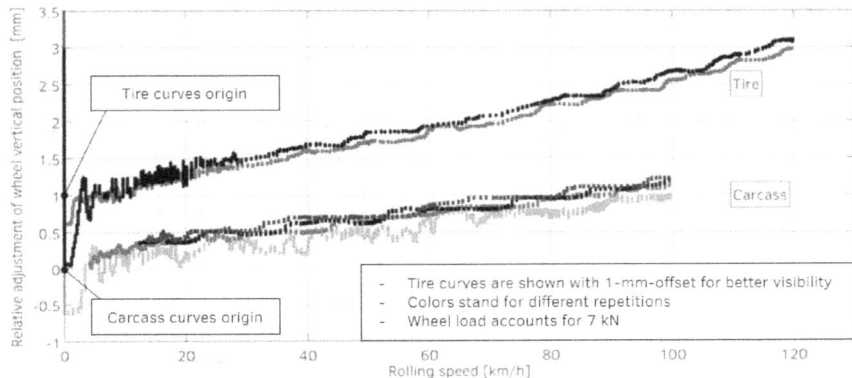

Figure 2.8. Relative adjustment of wheel vertical position depending upon the rolling speed, which maintained the wheel load at the constant level.

The mentioned rate was observed for all wheel load values (3, 5, 7 kN), with both camber values of 0° and −4°. The same effect was found by the cornering wheel: here, the tire was rolling with a 3° slip angle, carcass with 1° (Appendix A.3-A.18). Each datasheet in the Appendix depicts the conditional length values and shows the histograms considering all revolutions, which were made with given wheel load. All three repetitions of the test were considered.

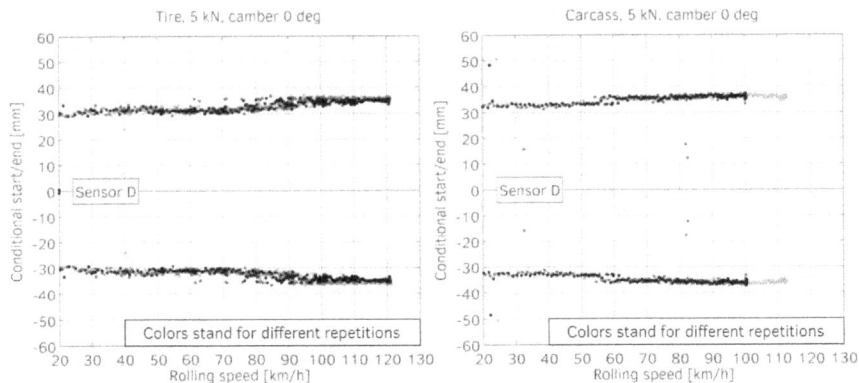

Figure 2.9. Conditional start and end of the contact patch depending upon rolling speed.

As a reference, the results of similar analysis with in-tire strain sensor can be used [Kim15]. The strain curve within one revolution described a similar tire curvature in [Kim15] as acceleration sensor did in this research. Variation of rolling speed also showed no significant difference in the contact patch length. This correspondence ensures the quality of the gained observations.

2.5.4 Contact patch geometry depending upon camber

Camber angle variation in the range of ±6° caused linear change in contact patch length, both for tire and carcass, for all three values of wheel load (Figure 2.10).

Test type:	Camber angle variation
Test object:	Tire
Wheel load:	3 – 5 – 7 kN
Rolling speed:	60 km/h
Slip angle:	0 deg
Camber angle:	-6 – +6 deg
Repetition:	first
	second
	third

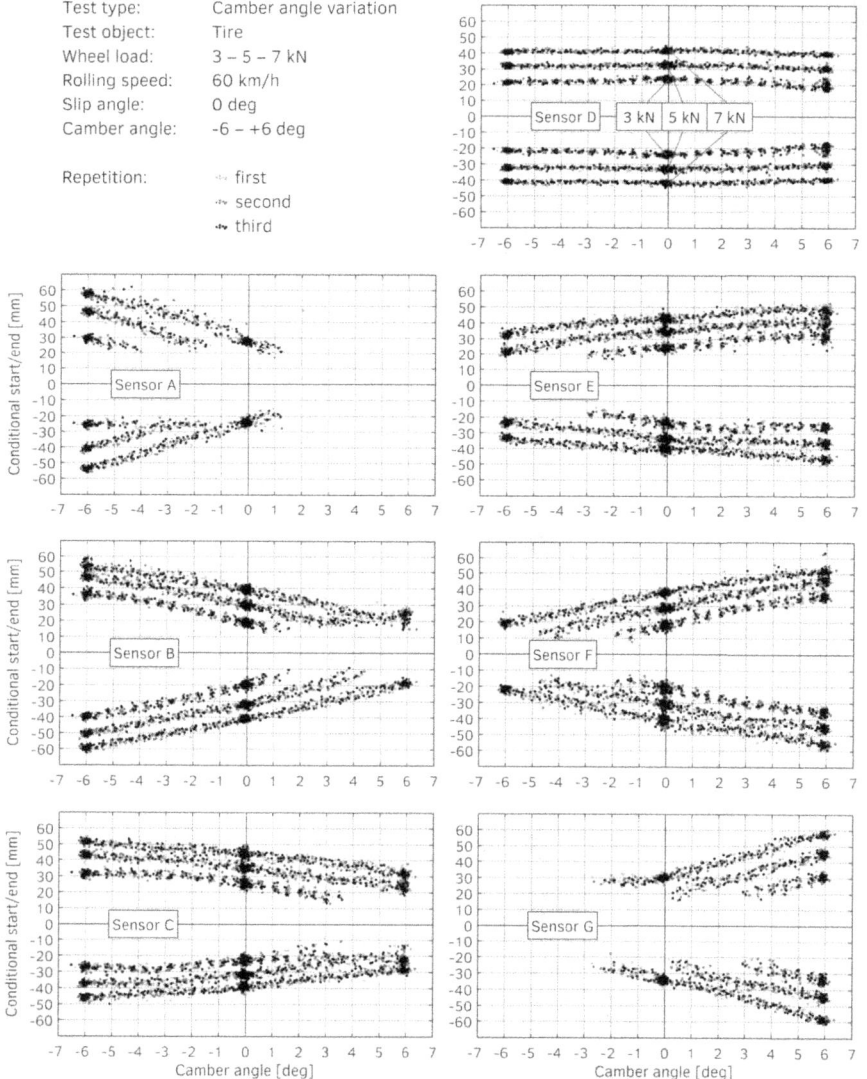

Figure 2.10. Positions of conditional contact patch start and end points during the camber variation for three values of wheel load (3, 5, 7 kN).

The rate of the length change achieved 25 % per degree of camber in sensors A and G, and was zero in the middle sensor D. The same measurement of the tire carcass showed qualitatively similar processes (Appendix A.19-A.20). Hence, the influence of tread on the contact patch shape of cambered tire is low.

2.5.5 Contact patch geometry depending upon slip angle

The first focus of this analysis was on contact patch change in the linear range of slip angle (±3°).

By means of FE-methods, it was shown in [Höf01, Kal10, Gar16] that the contact patch shape change is connected with lateral and longitudinal deformation of a tire. Taking into account tire structure, it is reasonable to conclude that the tire contact patch shape is mainly determined by carcass deformation. However, a cornering tire features both carcass lateral deformation and tread shear. For the purpose of the separation of these effects, the first measurement series was performed with the tire carcass sample (Figure 2.11). This measurement visualized the saturation of carcass lateral deformation (and consequently lateral force): The linear range for the carcass was only ±1°, and starting from 2.5° the length values received practically no further change. The linear range was investigated using a separated measurement series (Appendix A.23). Its results showed linear change of the contact patch length with the slip angle. Hence, the most possible reason for the observed behavior is linear connection between contact patch length and tire lateral deformation.

The same test of the complete tire in the linear range showed qualitatively similar results. However, the difference of the contact patch length was lower, because the carcass deformation was smaller (Figure 2.12).

In order to evaluate contact patch shape, a polynomial curve (order 3 to 5) was interpolated through seven measured points (Figure 2.13).

These observations led to two conclusions, which are valid for the given tire, testing conditions and three different wheel load values:

- The contact patch shape changes with slip angle in an asymmetric manner: the leading and trailing edges differ significantly (unlike the case of camber variation).

- The contact patch shape changes in a non-linear manner in regard to slip angle (again, unlike the situation with camber).

Numerically considered, the change rate of the contact patch length achieved 15 % per 1° of slip angle for the tire and 25 % per 1° of slip angle for the carcass.

The second focus of this analysis was contact patch shape in case of high slip angle, up to ±12°. These measurements showed much higher levels of oscillations of the tire body, so that the values of the conditional start and end position have much higher scatter. Still, these results confirmed the asymmetry of the contact patch shape. The comparison of contact patch length for different rolling speed values (60-90 km/h) revealed no significant difference (Figure 2.14, Appendix A.25-A.26).

Further measurements of cornering tire in the non-linear range (up to 12°) with the wheel load of 7 kN or for a cambered tire were disturbed with oscillations to an even greater extent. Due to the excessively high dispersion observed, these data could not be reliably analyzed (Appendix A.24-A.29).

Test type:	Slip angle variation (lin. range)
Test object:	Carcass
Wheel load:	3 – 5 – 7 kN
Rolling speed:	60 km/h
Slip angle:	-3 – +3 deg
Camber angle:	0 deg
Repetition:	first
	second
	third

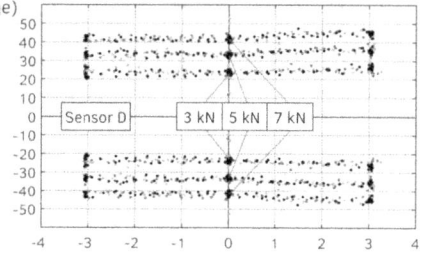

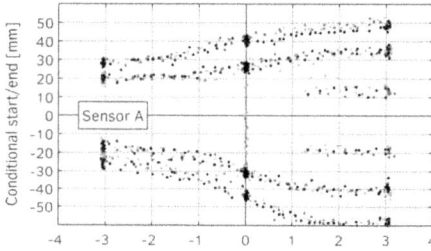

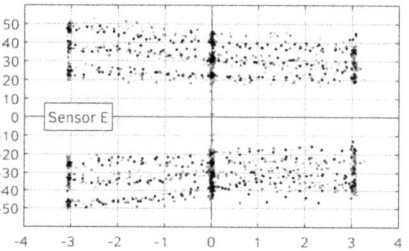

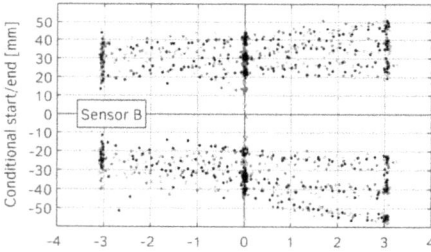

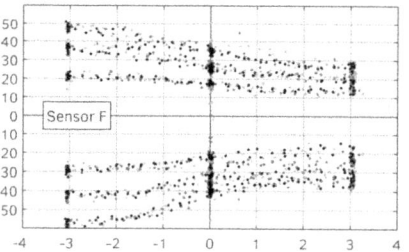

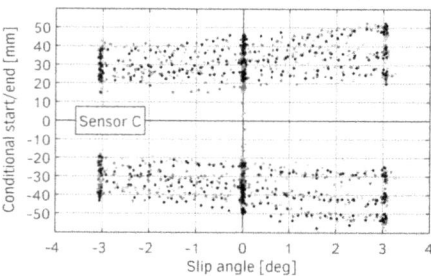

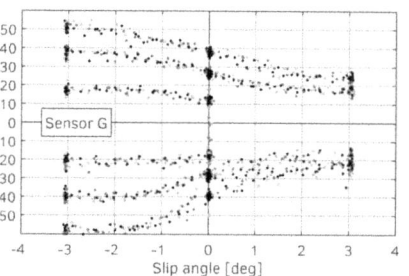

Figure 2.11. Positions of conditional contact patch start and end points of the carcass during the slip angle variation for three values of wheel load (3, 5, 7 kN).

Test type: Slip angle variation (lin. range)
Test object: Tire
Wheel load: 3 – 5 – 7 kN
Rolling speed: 60 km/h
Slip angle: -3 – +3 deg
Camber angle: 0 deg

Repetition: ⁓ first
                    ⁓⌄ second
                    ⁓⌄ third

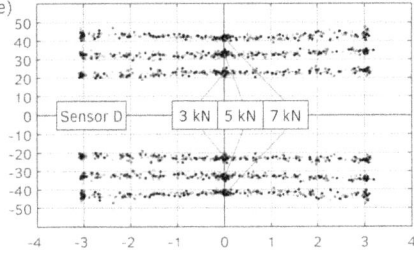

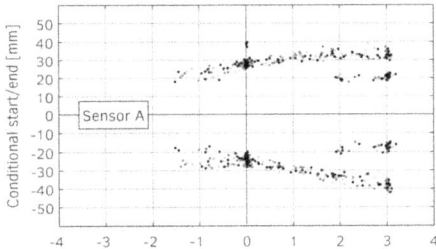

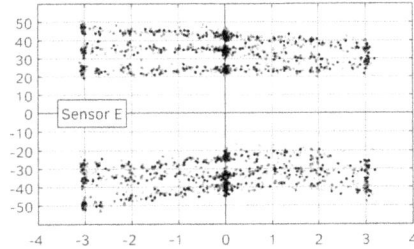

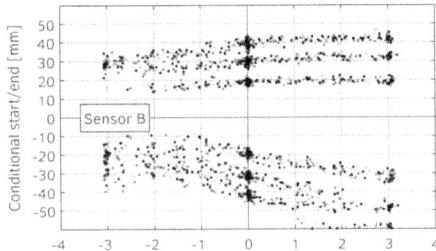

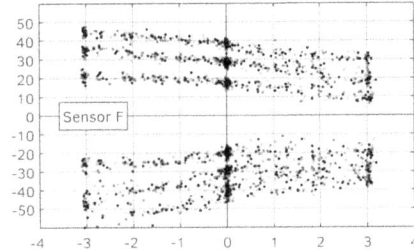

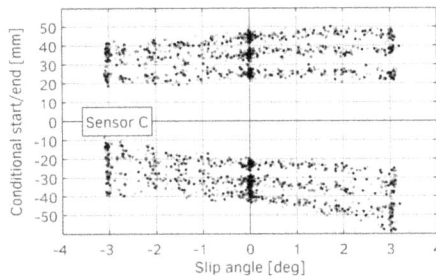

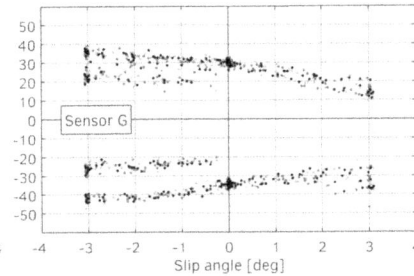

Figure 2.12. Positions of conditional contact patch start and end points of the tire during the slip angle variation for three values of wheel load (3, 5, 7 kN).

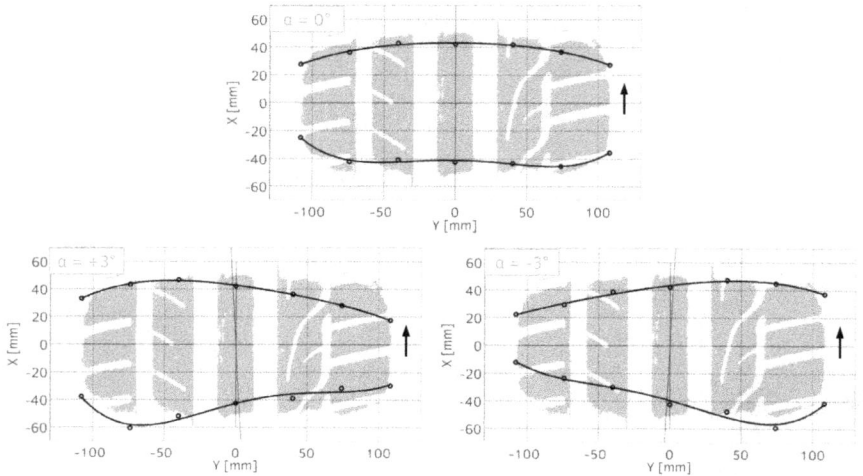

Figure 2.13. Interpolated leading and trailing edges of the contact patch at the wheel load of 7 kN.

Test type:	Slip angle variation	Repetition:	first
Test object:	Tire		second
Wheel load:	5 kN		third
Slip angle:	-1 – +12 deg		
Camber angle:	0 deg		

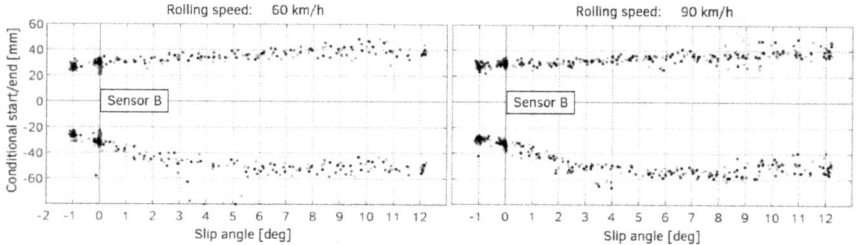

Figure 2.14. Positions of the conditional start and end of the contact patch for cornering tire with different rolling speed values.

2.5.6 Summary

The developed measurement method clarified how the contact patch shape of the specific tire changes depending upon slip angle, camber, rolling speed and wheel load. The physical background of the observed effects was not considered in this research. The goal was rather to observe them and to estimate, how relevant they are for tire handling properties. For this task, it was sufficient to describe the effects empirically. An overview of the observations is presented in Table 2.2.

Additionally, it was observed in all the conducted measurements that the contact patch length in the tire middle plane (sensor D) was insensitive to speed, camber or slip angle variation.

Table 2.2. Contact patch length behavior under different forms of excitation.

Test mode	Rolling conditions	Contact patch length of:	
		Tire	Carcass
Rolling speed variation 20 – 120 km/h	Free wheel $\alpha = 0°, \gamma = 0°$	- No significant change (+1.5 % per 10 km/h)	- No significant change (+1.5 % per 10 km/h)
	Cambered wheel $\alpha = 0°, \gamma = -4°$	- No significant change (+1.5 % per 10 km/h)	- No significant change (+1.5 % per 10 km/h)
	Cornering wheel $\alpha = 3°, \gamma = 0°$	- No significant change (+1.5 % per 10 km/h)	–
	Cornering wheel $\alpha = 1°, \gamma = 0°$	–	- No significant change (+1.5 % per 10 km/h)
Camber variation $\gamma = -6° ... +6°$	Free wheel $\alpha = 0°$, 60 km/h	- Linear change (up to 25 % per 1°) - Symmetric shape	- Linear change (up to 25 % per 1°) - Symmetric shape
Slip angle variation	$\alpha = -3° ... +3°$ $\gamma = 0°$, 60 km/h	- Slightly non-linear change (up to 15 % per 1°)	- Non-linear change (up to 25 % per 1°)
	$\alpha = -1° ... +1°$ $\gamma = 0°$, 60 km/h	–	- Linear change (up to 25 % per 1°)
	$\alpha = 0° ... +12°$ $\gamma = 0°$, 60 km/h	- Non-linear change - No reliable data after 5° - No significant influence of rolling speed - Asymmetric shape	–
	$\alpha = 0° ... +12°$ $\gamma = 0°$, 90 km/h		–
	$\alpha = 0° ... +12°$ $\gamma = 0°$, 60 km/h		–
Wheel load variation 3 – 7 kN	Free wheel $\alpha = 0°$ $\gamma = 0°$, 60 km/h	- Close to linear change (+14 % per 1 kN within the range 3 – 7 kN)	- Close to linear change (+14 % per 1 kN within the range 3 – 7 kN)

The conclusions from the contact patch shape investigation that contribute to achieving the set goal (model-based understanding of tire physics in handling) are following:

! The change of the contact patch shape depending upon the rolling speed can be neglected.

! The contact patch shape changes significantly with slip angle, this must be considered.

! The change of the carcass contact patch shape gives better insight into tire contact patch shape variation: The non-linear curve is straightened because of tread deformation. Hence, although not visually noticeable, the non-linearity of the change in tire contact patch length depending upon the slip angle must be taken into account.

! The contact patch length changes with wheel load significantly in a close to linear manner, this must be considered.

With these four statements and the numerical data behind them (Table 2.2), the effect of tire contact patch shape change can be introduced into the simulation model in order to clarify how does it influence tire handling behavior. Next important issue is elastic behavior of tire carcass.

2.6 Tire carcass deformation

2.6.1 Motivation and measurement concept

The physical background of tire lateral force is the lateral deformation of its flexible body. Hence, in order to understand transient lateral force generation, tire lateral deformation must be understood. The first challenge is that the lateral tire deformation is a sum of its carcass deflection and tread shear. The tread in this sense is approximately one order of magnitude stiffer than the carcass (Figure 2.15). The entire deformation of a tire can be calculated based on rolling kinematics; however, it remains unknown, which part of it is carcass deflection and which part is tread shear. Consequently, it is necessary to measure at least one of these two deformation components.

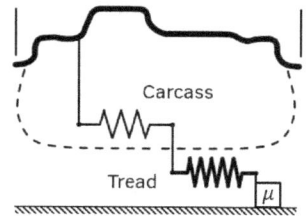

Figure 2.15. Lateral deformation of a tire: Two flexible elements and a frictional connection to the road surface.

The second difficulty is that lateral carcass deformation is a composition of a number of different effects, such as lateral translational deflection, lateral bending, and rotation around a radial axis. To understand these, it is necessary to observe the whole circumference of the carcass body or at least the whole circumference of the carcass centerline (equator).

Based on two mentioned requirements, a novel method to investigate carcass deformation behavior was developed. A rim with eight windows (Figure 2.16a) was designed. Inside the rim, cameras were mounted, one per window (Figure 2.16b). On the inner surface of the tire carcass, markers were glued (Figure 2.16c). The markers were white reflective dots with diameter below 1.5 mm on square black stickers.

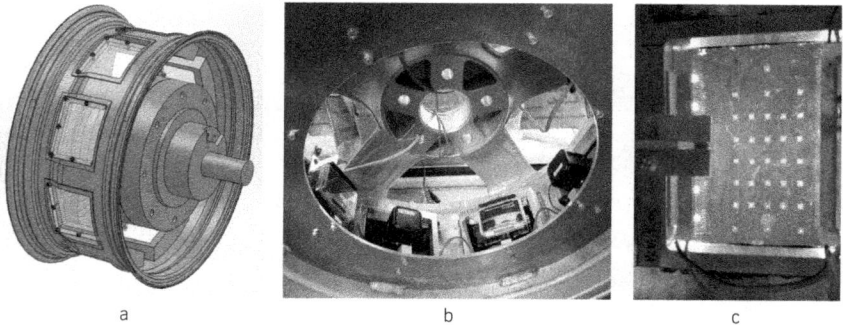

a   b   c

Figure 2.16. "Looking inside the rolling tire": The concept of optical measurement of carcass deformation behavior.

The developed system has made it possible to perform measurements focused on various issues, which will be described in separate subchapters.

2.6.2 Lateral stiffness of tire carcass

As lateral deformation of the tire plays a significant role in lateral force and aligning torque generation, it is important to understand the connection between deformation and force, namely stiffness and damping. For this purpose, there were performed measurements of tire lateral stiffness. The tire was stationary on the flat plate. The plate moved in lateral direction at different speed values. The eight cameras made it possible to follow the deflection of the entire carcass circumference.

In order to avoid influence of inertia force, tire was excited with plate displacement with constant velocity (instead of sinusoidal excitation). Damping properties are investigated in this research only to explore limits of the damping force. For this reason, measurements were conducted with only three values of excitation velocity (1, 100 and 200 mm/s, Figure 2.17) and evaluated in time domain. Because of limitations of testing equipment, the set velocities of 100 and 200 mm/s were not achieved properly (Figure 2.17e-f), but this excitation was enough to estimate the limits of the damping force. Detailed analysis of the damping properties is a subject for further research.

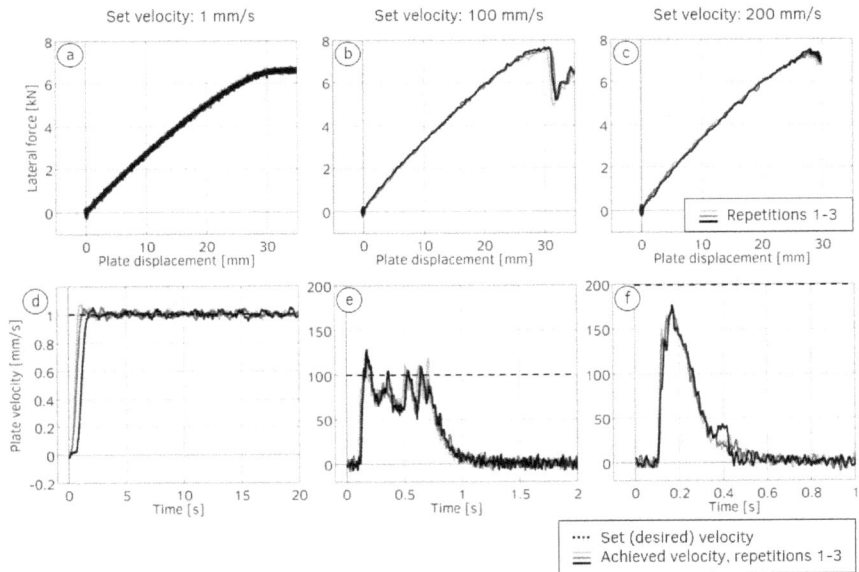

Figure 2.17. Tire lateral stiffness depending upon the speed of lateral deformation.

The development of lateral force with plate displacement revealed three phases (Figure 2.18):

1. The linear area, in which there is practically no sliding yet. The plate displacement is the sum of carcass and tread deformation.
2. The non-linear area, in which some tread elements stick and some slide. This region can also include a decrease of the force with increasing displacement due to the Stribeck-Effect and frequency-dependent friction [Sel14, Tor15, Kel12].
3. The horizontal area, in which all tread elements in the contact patch slide.

The analysis of the linear area has the advantage that the friction properties have practically no influence on it. Hence, the carcass stiffness and damping could be investigated in this area directly.

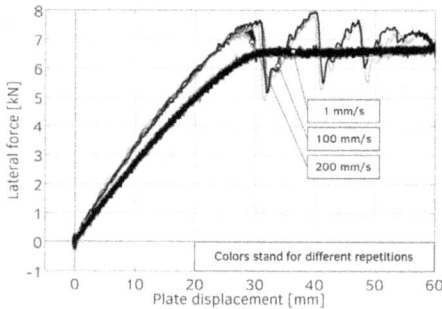

Figure 2.18. Force-displacement diagram for tire lateral translational excitation.

The comparison of linear slopes of the force-displacement diagrams (Figure 2.18) showed significant difference between measurements at 1 mm/s and 100 mm/s; at the same time, the slopes measured at 100 mm/s and 200 mm/s were almost identical. Hence, the flexible body of the tire did exhibit frequency-dependent properties in the direction of lateral deformation within the linear range, namely damping.

This behavior may be caused by both carcass and tread, because tire lateral flexibility is flexibility of two spring elements connected in a series (carcass and tread, Figure 2.15). The tread shear modulus of a passenger car tire is frequency-dependent [Soc05, Kel12]. In order to investigate limits of carcass damping properties, carcass deflection was captured optically (Figure 2.16) for three mentioned measurements of tire lateral stiffness.

Each of eight cameras captured coordinates of corresponding group of reflective markers, glued on the tire inner liner (Figure 2.19). In each group, there was selected one central marker. The central marker 1 was placed in the center of the contact patch. Hence, camera 1 was oriented vertically downwards. Measurement results are shown as inverted force-displacement diagrams, which depict lateral displacement of the plate and the central markers in relation to the tire lateral force (Figures 2.19-2.20).

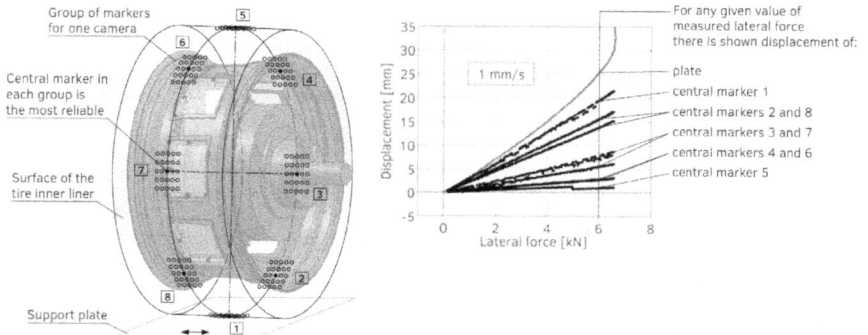

Figure 2.19. Carcass displacement investigation by means of tire lateral stiffness measurement.

Quasi-static deformation (Figure 2.20a) showed a linear connection between the carcass displacement and the lateral force. Consequently, for a given tire and in a working range of the lateral deformation:

! The lateral stiffness of tire carcass can be considered independent of the displacement.

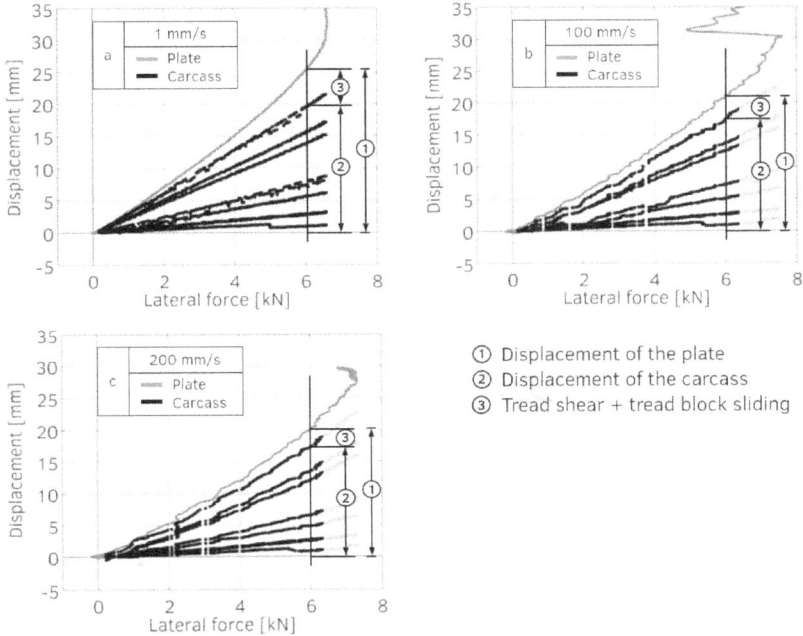

Figure 2.20. Displacement of the central markers on the carcass in relation to the lateral force.

Difference between plate displacement and carcass deflection (dimension 3, Figure 2.20) in the non-linear range (15-30 mm) stands for the sum of tread shear and tread block sliding, including lateral overrolling of the tire. The higher the plate speed, the smaller the difference. A physical explanation of the effect is frequency-dependent friction. According to Figure 2.18, for the given tire and the given road surface (3M Safetywalk Ref. 610), slide friction coefficient accounts approximately for 0.93 at 1 mm/s sliding speed, and for 1.08 at 100 mm/s. Hence, higher sliding speed leads to higher friction force. A consequence of higher force is higher deflection of the tire carcass. This logical chain confirms that the analysis of the carcass deflection is qualitatively correct.

Figure 2.20 shows that the linearity of the carcass displacement curve varies with the deformation speed. To analyze this difference, measurements of carcass displacement in the central marker 1 were evaluated. This marker is in the center of the contact patch. Hence, it is the last point to start sliding. These three curves were compared in a traditional (not inverted) force-displacement diagram (Figure 2.21). With help of the trend line, it can be seen that low-speed excitation caused a linear force response to the carcass deflection. For high-speed excitation, the same carcass deformation (e.g., 10 mm) caused a perceptibly higher force (3.70 kN to 3.15 kN). The slope remained generally the same, so the stiffness properties were the same. Therefore, for the given tire, wheel load and inflation pressure:

> Lateral damping force of the tire carcass varies from 0 kN to 0.55 kN within the working range of lateral velocities. The connection between the velocity and the damping force is not linear.

Analysis of measurement results for three different wheel loads proved the reproducibility of the described effects (Appendix A.30-A.32). Furthermore, it was revealed that the tread deformation decreases with wheel load growth (Figure 2.22). This occurred because with the higher wheel load, more tread elements belonged to the contact patch. Therefore the stiffness of the tread layer increased (Table 2.3).

General limitation of this method is optical distortion of marker recognition procedure because of tire radial deflection. Although a correction of this effect was performed based on difference in marker positions before and after applying the wheel load, numerical comparison cannot provide an accuracy better than 15 % error.

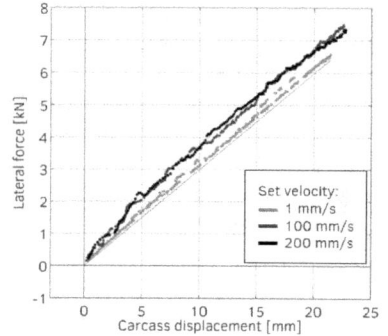

Figure 2.21. Comparison of connection between carcass displacement and lateral force for three set velocities of the plate.

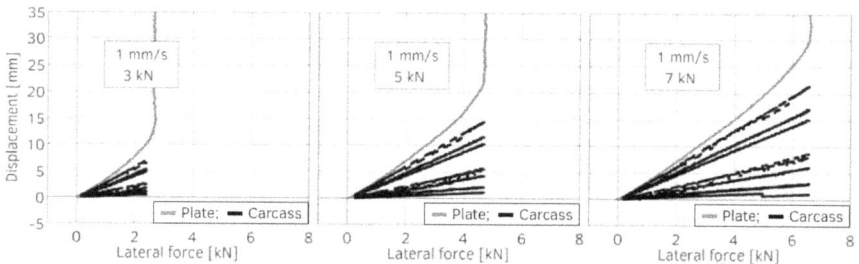

Figure 2.22. Comparison of displacement of plate and carcass for three different wheel load values.

Table 2.3. Contact patch dimension depending upon wheel load.

Wheel load	[kN]	3 kN	5 kN	7 kN
Contact patch length (on flat surface)	[mm]	70	90	120
Contact patch width (on flat surface)	[mm]	170	180	200
Linearized stiffness of the total carcass (displacement according to the center marker 1)	[kN/mm]	0.320	0.309	0.299

The measured slope of the lateral force depending upon the carcass displacement slightly decreased with the wheel load growth (Figure 2.23, Table 2.3). However, taking into account the mentioned problem, this difference should be investigated with higher precision.

The difference in the contact patch length does not affect carcass lateral stiffness significantly, because the contact patch length accounts approximately to 5 % of tire circular length. This conclusion was confirmed with help of direct measurement of the carcass focused on its lateral stiffness (Figure 2.24c). Figure 2.24 shows force-displacement diagrams for carcass (a) and tire (b) measured at 1 mm/s for three wheel load values. According to Figure 2.24a, the carcass stiffness in the linear range was independent upon wheel load and consequently upon contact patch length.

Figure 2.25 shows that the difference between tire-curve and carcass-curve in the linear range for 7 kN was noticeably lower than for 5 kN, and much lower than for 3 kN. This means that tread deformation is lower with higher wheel load, because more tread elements belong to contact patch.

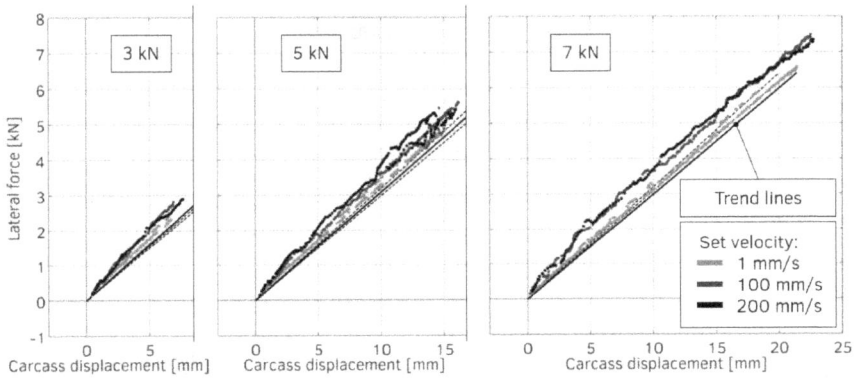

Figure 2.23. An overview of flexible and damping properties of the tire in the lateral direction.

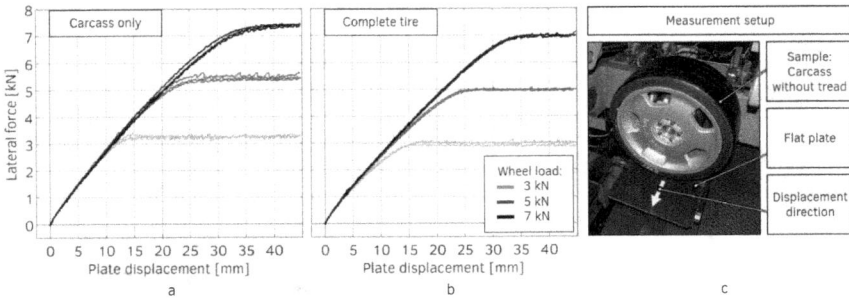

Figure 2.24. Lateral stiffness measurement of the carcass and the tire.

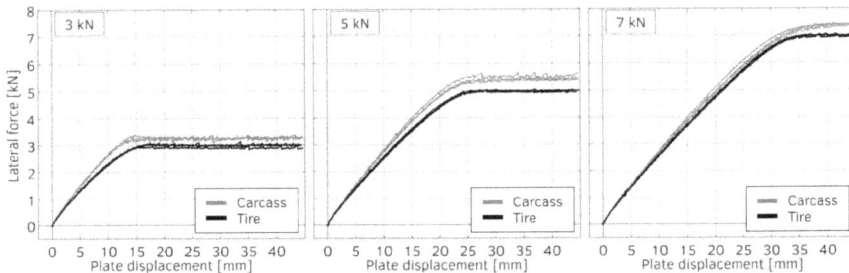

Figure 2.25. Comparison of force-displacement-diagrams of the carcass and the tire.

For the given tire in the wheel load range from 3 kN to 7 kN it is fair to conclude the following:

Carcass lateral force divided by integrated lateral deflection of its body delivers the distributed lateral stiffness of approximately 0.3 N/mm². With a higher wheel load:

- lateral stiffness of the tread layer <u>increases</u>, proportionally to the effective area of the contact patch (approx. 15 % per 1 kN);
- lateral stiffness of the carcass shows <u>no significant change</u>;
- lateral stiffness of the complete tire receives <u>no significant change</u>, namely due to the connection of carcass and tread in series (approx. 1.5 % per 1 kN).

With these insights, the analysis of lateral stiffness and damping of the carcass can be concluded. The next important deformation component of the carcass is bending.

2.6.3   Carcass bending properties

Tire carcass experiences bending in different directions, therefore it is necessary to define the terms involved in this research. In the sense of handling, namely in the generation of tire lateral force and aligning torque, carcass <u>lateral</u> flexibility is essential. This can be generally described with a physical model "string on an elastic foundation" (Figure 2.26, 2). To simplify this three-dimensional model to a flat model, the tire carcass can be hypothetically cut in its top point and unfolded onto the road surface (Figure 2.26, 3-4). In such a manner, the system is transformed into the typical case of the beam theory: a flat beam on an elastic foundation (Figure 2.26, 5).

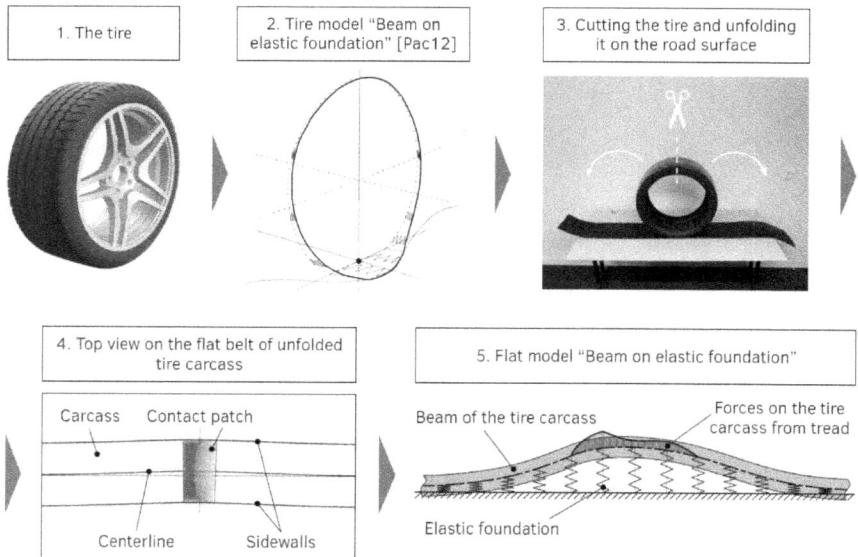

1. The tire	2. Tire model "Beam on elastic foundation" [Pac12]	3. Cutting the tire and unfolding it on the road surface

4. Top view on the flat belt of unfolded tire carcass	5. Flat model "Beam on elastic foundation"

Carcass   Contact patch

Centerline   Sidewalls

Beam of the tire carcass   Forces on the tire carcass from tread

Elastic foundation

Figure 2.26. Transformation of a three-dimensional system of tire to a two-dimensional beam on an elastic foundation.

Carcass bending – the deflection of a carcass body in its lateral direction under the forces and torques, applied in planes, tangential to the cylindrical body of the carcass: lateral force, aligning torque, carcass lateral elastic force.

In other words, it is the bending of the carcass around any radial axis of the wheel. This term is not to be confused with carcass bending due to vertical deflection of the tire.

The described bending contributes to the lateral deflection of tire carcass and significantly influences tread shear deformation. Hence, this effect is relevant for lateral force and aligning torque generation. To understand the bending behavior of a tire carcass, the following questions must be clarified:

?   How does the carcass centerline deform in the lateral direction? Which curve can describe its form (Figure 2.26, 5)?

?   What are numerical values of carcass bending stiffness in the lateral direction?

The optical measuring system described in subchapter 2.6.1 delivers the carcass deflection curve. The described measurement of tire lateral stiffness provides a form of deflected carcass, captured by eight cameras. Figure 2.27 illustrates, first of all, that the wheel load has no significant influence on tire lateral deflection in the linear range, as observed in subchapter 2.6.2. Secondly, the form of the deflected carcass in the contact patch (range ±100 mm) is close to parabolic. However, a parabola is unable to describe the form both inside and outside the contact patch.

Further insight into the deflected carcass can be provided by an optical measurement with the same camera-based system, but for the case of tire torsional excitation, also called parking test. In this test, a stationary tire is turned around the vertical axis.

D   **Bore torque** – the component of the aligning torque that is caused by non-zero yaw velocity of the tire and corresponding spin (pivoting) friction forces in the contact patch.

If the tire is not rolling, the bore torque is the only component of the aligning torque and consequently can be also named M_Z.

D   **Torsional stiffness of a tire** – the stiffness of a stationary tire in the yaw direction. This indicate the relationship between the bore torque and the angle of tire turning around its vertical axis.

This term was adopted in [Pac12] and is not to be confused with torsional stiffness in the pitch direction (tangential elasticity). The laterally-deflected carcass for this case is illustrated in Figure 2.28. These data confirm previous observation that the change of wheel load does not perceptibly influence the lateral deflection of the carcass.

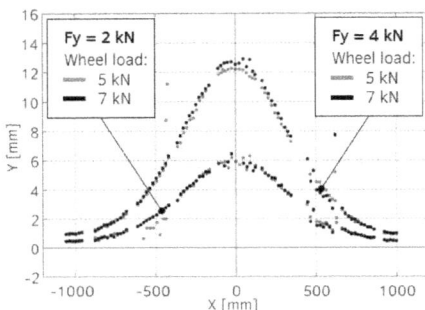

Figure 2.27. Lateral deflection of the carcass by the lateral stiffness measurement.

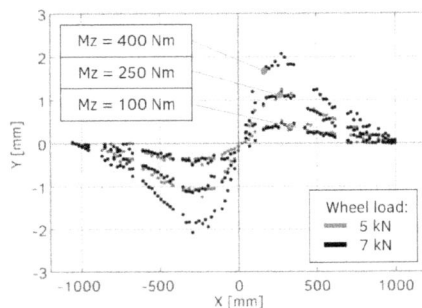

Figure 2.28. Lateral deflection of the carcass by the torsional stiffness measurement.

The measurement of the torsional stiffness of the stationary tire depicts the very low carcass lateral deflection (below 2 mm for the given conditions). Consequently, the significant part of the bore torque is caused by carcass bending stiffness. Lateral flexibility does not perceptibly influence it. Hence, by measuring the torsional stiffness with different yaw velocity values, it is possible to estimate the damping effects of carcass bending (Figure 2.29). Once again, desired yaw rotation speed was not achieved properly by the testing equipment (Figure 2.29b-c). Due to this, only rough estimation of the damping properties could be made, but it is enough to explore their limits.

Due to the high torsional stiffness of the carcass, the torque-angle diagram emphasizes the very small linear range: The significant sliding in the contact patch starts already by yaw angle of 1°. Within this range, there is practically no sliding between the tire and the plate. Independently upon different yaw rotation speed values and profiles (Figure 2.29), three torque-angle curves are very close to each other in the linear range (Figure 2.30). This is a first sign that the bending properties of the carcass do not exhibit significant damping.

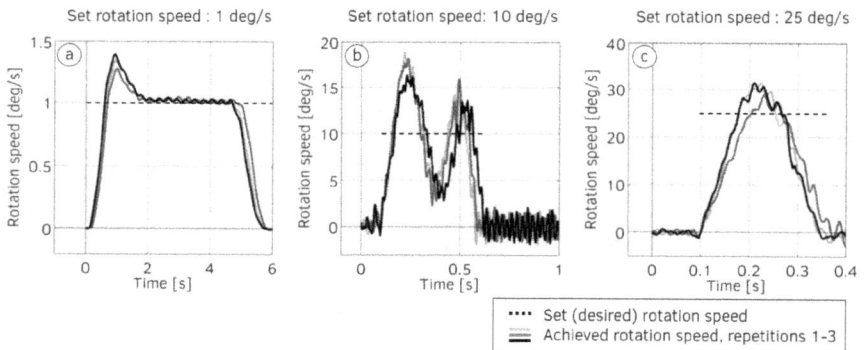

Figure 2.29. Yaw rotation speed profiles for three set values.

The difference of aligning torque generation outside the linear range with different yaw speed is perceptible (Figure 2.30): A given value of aligning torque (e.g., -0.4 kNm, Figure 2.30c) is achieved with different turning angles depending upon the speed.

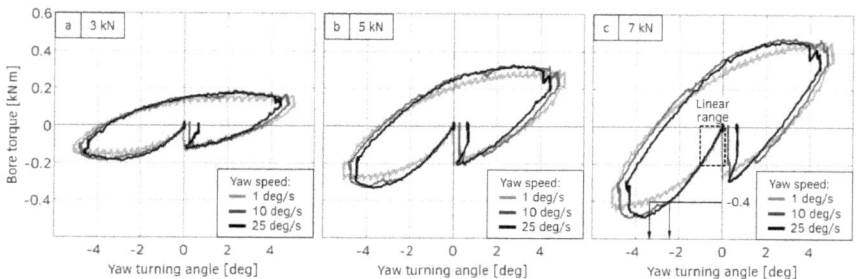

Figure 2.30. Bore torque generation with different yaw speed.

In order to clarify which of the tire properties contribute to this difference, carcass deflection was compared by the given bore torque threshold (–0.4 kNm, Figure 2.31).

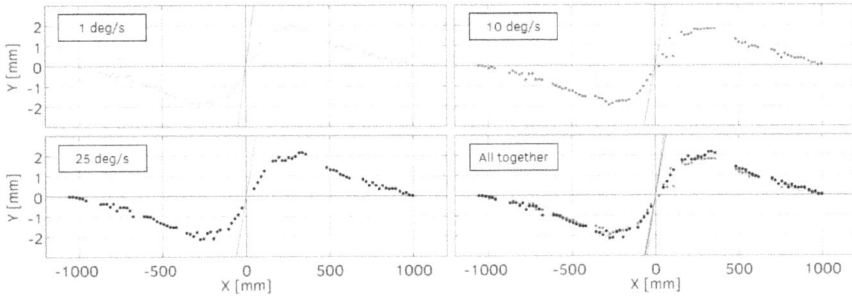

Figure 2.31. Lateral deflection of the tire carcass corresponding to the bore torque of 0.4 kNm depending upon the yaw turning speed.

The accuracy of the optical measuring concept with eight in-rim cameras did not allow a numerical statement to be made regarding differences in the lateral deflection curves (because of low deflection values, ca. 2 mm). However, a qualitative conclusion could be made: There was no significant difference between the three deflected forms of the carcass that corresponded to the same bore torque. Hence, for the given conditions:

!   The bending behavior of the carcass does not exhibit perceptible damping.

As with lateral stiffness (Figure 2.24), in order to understand the bending properties of the carcass, it is helpful to measure not only the complete tire, but also the carcass without tread (Figure 2.32c). These measurements showed higher slope in the torque-angle diagram of the carcass compared to the tire, because of the absence of the tread layer (Figure 2.32a-b; Figure 2.33).

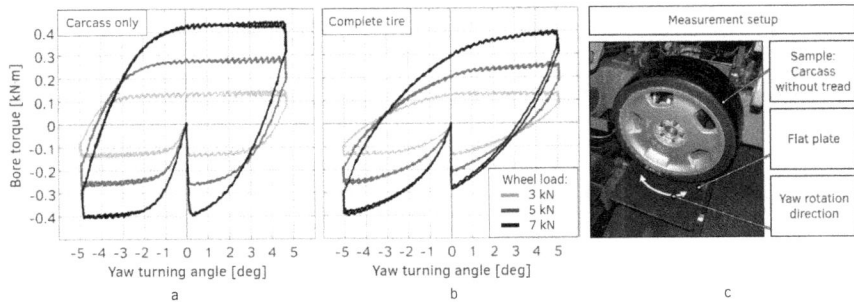

Figure 2.32. Torsional stiffness measurement of the carcass and the tire.

Interpretation of Figures 2.32-2.33 allows to summarize the following for the given conditions:

!   Torsional stiffness of the tire carcass in the linear range <u>does not change significantly</u> with the wheel load variation.

Background: contact patch length accounts for ca. 5 % of carcass circumference length.

!   Torsional stiffness of the complete tire in the linear range <u>does change perceptibly</u> with the wheel load variation.

Background: An increase of contact patch area significantly increases the polar sectional modulus of the tread layer in the contact patch, so the torsional stiffness of this layer becomes higher. In other words, the higher the wheel load, the closer the torque-angle curve of the tire to that of the carcass.

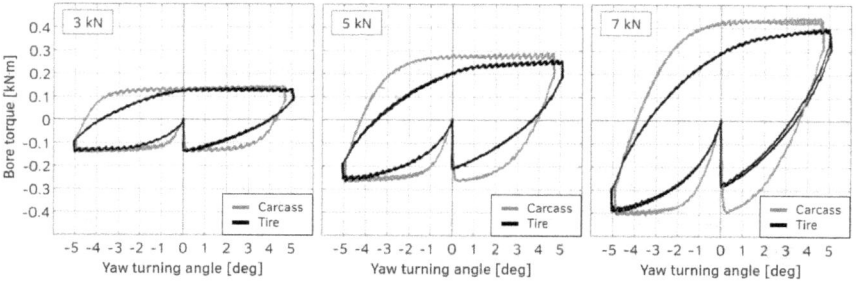

Figure 2.33. Comparison of torque-angle diagrams of the carcass and the tire.

In such a manner the stiffness and damping properties of the carcass lateral elasticity and lateral bending were investigated, and the deflected forms of its centerline (equator, neutral axis) were gained. However, knowing the bending of the neutral axis is insufficient to know the bending of the entire wide body of a carcass.

? How does the entire cylindrical body of a tire carcass deform?

Does it happen without rotation of its cross-sections (rotation-free hypothesis, Figure 2.34), or without shear between these cross-sections (shear-free, Euler-Bernoulli hypothesis), or with a combination of these effects?

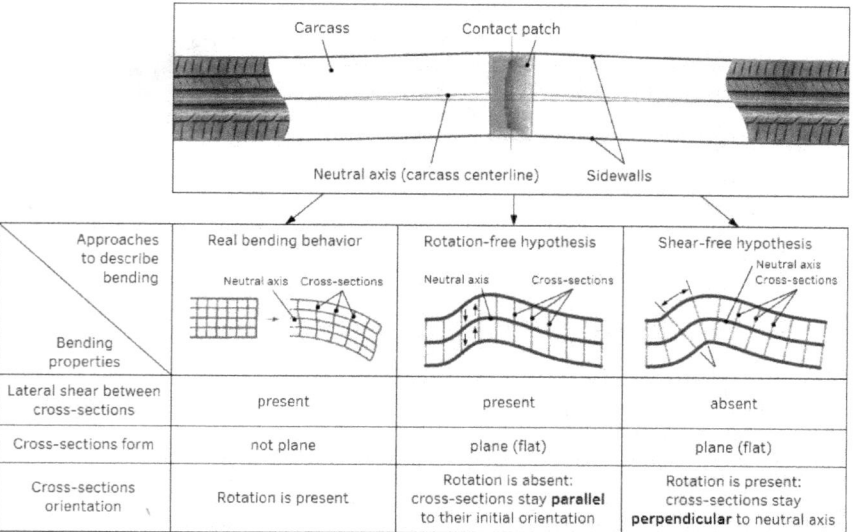

Approaches to describe bending / Bending properties	Real bending behavior	Rotation-free hypothesis	Shear-free hypothesis
Lateral shear between cross-sections	present	present	absent
Cross-sections form	not plane	plane (flat)	plane (flat)
Cross-sections orientation	Rotation is present	Rotation is absent: cross-sections stay **parallel** to their initial orientation	Rotation is present: cross-sections stay **perpendicular** to neutral axis

Figure 2.34. Different approaches to describe carcass bending behavior.

To answer this question, a stereocamera-based optical system was used (Figure 2.35). Two cameras captured the tire on the test rig with a frequency of 15 Hz. Based on a pair of images, the three coordinates of the markers on the tire were defined with a precision of 0.01 mm.

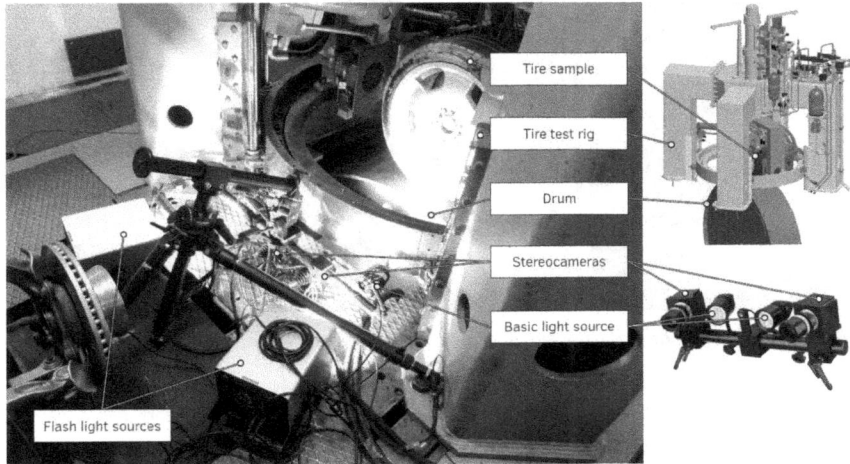

Figure 2.35. The measurement setup of the stereocamera-based system.

As the goal was to capture the deflection of the cylindrical part of the carcass, the shoulder edge of the tire was used. The tread blocks on the shoulder were cut in such a manner that each block provided a flat surface perpendicular to the wheel rotation axis (Figure 2.36a). The depth of this cutout was selected considering the width of the contact patch: The edge was not allowed to enter the contact patch even in the cornering tire with high wheel load (Figure 2.36b). Markers were glued to the flat surface on each block. The second array of markers was glued to the rim (Figure 2.36c). Assuming the rim to be a rigid body, these markers were used to capture the wheel rotation angle and slip angle, or the yaw angle in cases of torsional stiffness measurement.

The described measurement method provided two important insights:

- The bending behavior of the carcass;
- The lateral deflection of the carcass, measured on its shoulder (not on the equator).

Regarding the first insight, the distance between the pairs of markers was analyzed using the measured data (Figure 2.37).

> **D** **Cross-section of the carcass in any given marker** is a plane, which passes through this marker and is perpendicular to the neutral axis of the carcass in unloaded condition (slip angle is zero).

In unloaded condition of the carcass, the distance between two cross-sections on the neutral axis and on the shoulder were the same (l), because the cross-sections were flat and parallel to each other.

As soon as deformation occurred (non-zero slip angle), the distance between these two cross-sections on the neutral axis remains the same (l). However, the distance between these cross-sections on the shoulder changed to $l + dl$. This consideration is still simplified compared to reality.

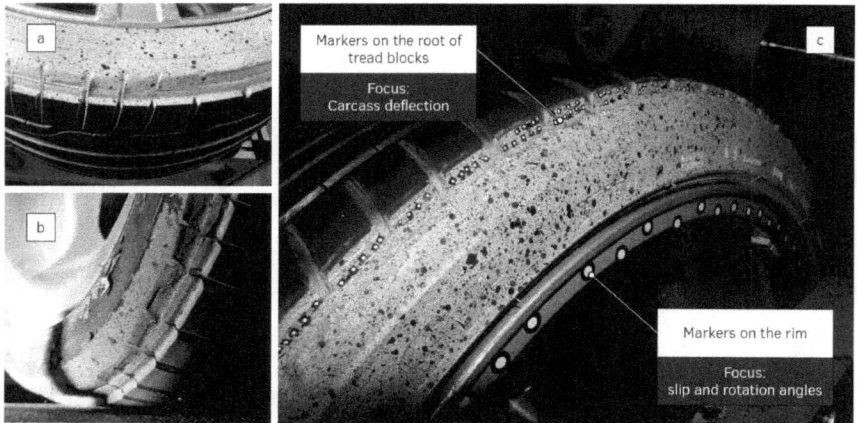

Figure 2.36. Preparation of the tire and positioning the markers on the tire and the rim.

Knowing the distance l, the difference dl, the width between two shoulder edges and the form of the neutral axis, it was possible to compare the bending behavior with two boundary cases, namely shear-free and rotation-free hypotheses.

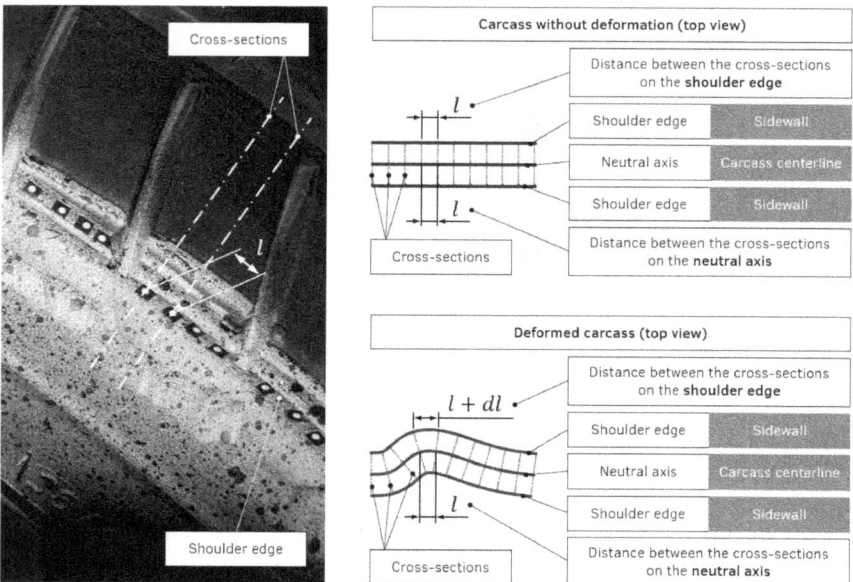

Figure 2.37. An approach to estimate carcass bending behavior based on deflection of its shoulder.

The first experiment with the stereocamera system corresponded to the measurement of tire torsional stiffness, previously performed with eight in-rim cameras. Here, the stationary wheel was turned around its vertical axis from 0° to 5° of yaw angle. The field of view of the two cameras covered 15 tread blocks (Figure 2.38a). Two reliable markers were selected on each block. Their lateral deflection in relation to their initial position in the wheel coordinate system is shown in Figure 2.38b.

The distance between each pair of markers $l + dl$ was evaluated depending upon the yaw angle change. The distance was normalized as $(l + dl)/l$. Figure 2.38c shows values of this normalized distance for corresponding tread blocks. The different symbols included in this diagram represent different repetitions of the measurement. Two curves stand for approximation of the received scatters.

For several blocks $(1, 6, 10, 15)$, normalized distance behavior in time domain is shown (Figure 2.38d). The dashed line depicts the scaled curve of the yaw angle.

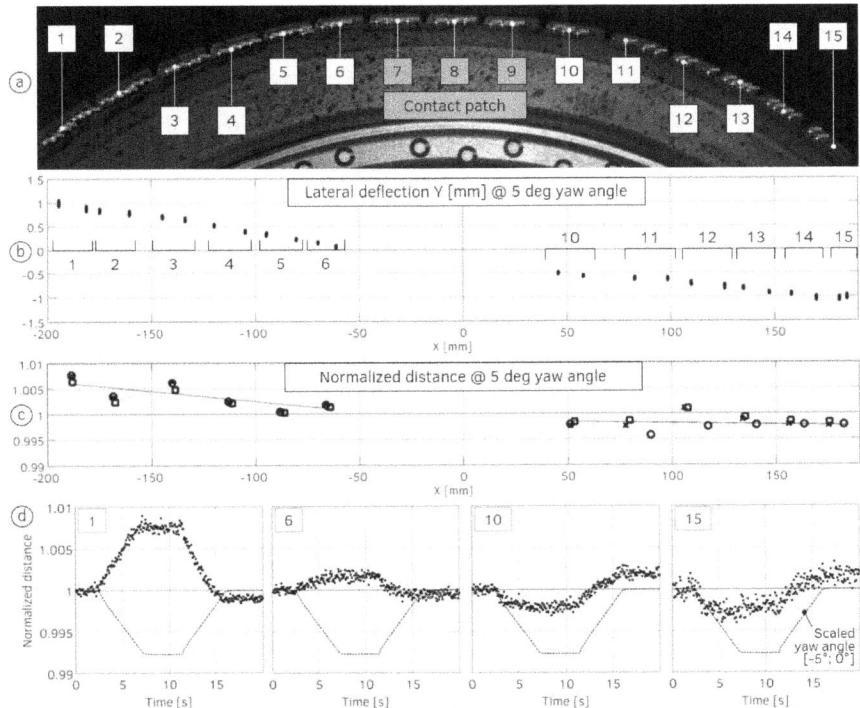

Figure 2.38. Analysis of the carcass deformation behavior based on the distance between the markers on the carcass shoulder.

This analysis delivered a qualitatively correct picture. The markers on blocks 1-6 moved away from each other (the normalized distance increased with the growing yaw angle, namely by 0.2-0.7 %). The markers on blocks 10-15 moved towards each other (the normalized distance reduced, namely by $-0.3 \dots -0.2$ %).

In order to be able to capture the deformation behavior, higher deformation of the tire was required. This occurred during the cornering, when the slip angle varied by high values outside the linear range of cornering stiffness (Figure 2.39).

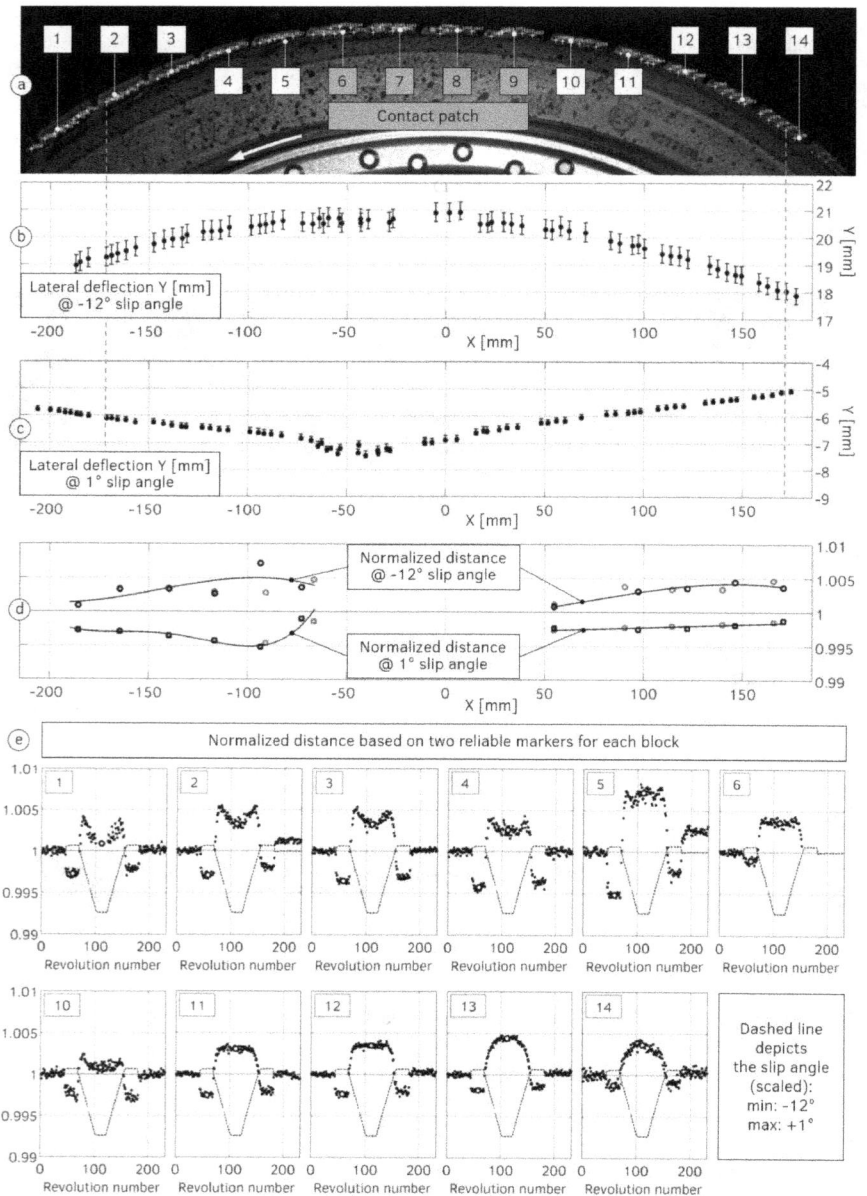

Figure 2.39. Analysis of the carcass deformation behavior during cornering with 7 kN wheel load.

The following measurement procedure was used: The tire was rolling at 60 km/h, the wheel load was 7 kN, the slip angle was changed from 0° to 1°, then to −12°, then back to 1° and finally back to 0°.

The stereocameras captured a frame once per revolution, providing the coordinates of the markers. Figure 2.39a shows the tire upside down, as it was captured, with the numbers of the blocks.

Figure 2.39b depicts the lateral deflection of the markers captured at a slip angle of −12°. These deflection values were measured in relation to markers initial position in the wheel coordinate system.

Figure 2.39c shows the same deflection, but captured with 1° of slip angle.

Figure 2.39d depicts the values of normalized distances between two selected markers of each block, namely for −12° and 1° of slip angle. The different symbols in this diagram stand for different repetitions of the measurement. Four curves stand for approximation of the received scatters.

Each of the 11 diagrams in Figure 2.39e represents the normalized distance between two markers for each block (except for blocks in the contact patch).

In Figure 2.40, normalized distances are depicted for two blocks (one before and one after the contact patch). The lateral deflection values of these blocks as well as the normalized curves of the aligning torque and lateral force measured during the experiment are also shown in Figure 2.40. The dashed line in each of these four diagrams shows the scaled curve of the slip angle. Analysis of these data can be structured using several important observations.

Observation 1: Lateral deflection of the carcass at a slip angle of −12° is symmetrical. The entire contact patch slid in the lateral direction (except for a very short area on the leading edge).

Observation 2: Lateral deflection of the carcass at a slip angle of 1° depicted a close-to-linear form in the contact patch and before it. The high curvature of this deflected line at the end of the contact patch was caused by a second row of the markers, which were placed closer to the road surface (Figure 2.36) and therefore showed also shear deformation of the tread blocks.

Observation 3: Within the linear range of the cornering stiffness, the normalized distance changed linearly with the slip angle. This was valid for all blocks (Figure 2.40a-b).

Observation 4: The normalized distance between the two markers on each block reproducibly changed with the slip angle. The form of this connection changed with a continuous trend from block to block.

Observation 5: For the entire range (both linear and nonlinear, 0°-12°), the behavior of the various blocks differed: For the blocks before the contact patch (blocks 10-14), the normalized distance increased monotonously with the slip angle (Figure 2.40b). For the blocks after the contact patch (blocks 1-6), the normalized distance changed non-monotonously with the slip angle (Figure 2.40a): As soon as the linear range of the cornering stiffness ended, it reduced significantly.

Observation 6: The curve of the normalized distances before the contact patch (Figure 2.40a) qualitatively correlated with the aligning torque curve (Figure 2.40e). This confirmed that the leading part of the tire is essential for aligning torque generation beyond the linear range, because the carcass deflection curvature in the trailing part is already saturated in this range.

Observation 7: The curves of the lateral deflection (Figure 2.40c-d) qualitatively correlated with the lateral force (Figure 2.40f). This complies with the linear connection between carcass lateral deflection and tire lateral force, which was observed in subchapter 2.6.2.

<u>Observation 8:</u> When the slip angle equals 1°, a reduction of the normalized distance of the blocks <u>before</u> the contact patch (blocks 10-14) had a low value (ca. −0.25 %). This reduction value changed from −0.12 % for block 14 to −0.27 % for block 10, linearly (Figure 2.39d). The reduction of the normalized distance of the blocks <u>after</u> the contact patch (blocks 1-6) had a higher value and followed a nonlinear trend from block to block: Starting with −0.12 % for block 6 (trailing edge), it increased to −0.50 % for block 5 and then decreased to −0.27 % for block 1. This figure correlates qualitatively with the deflected curve of the carcass (Figure 2.39c).

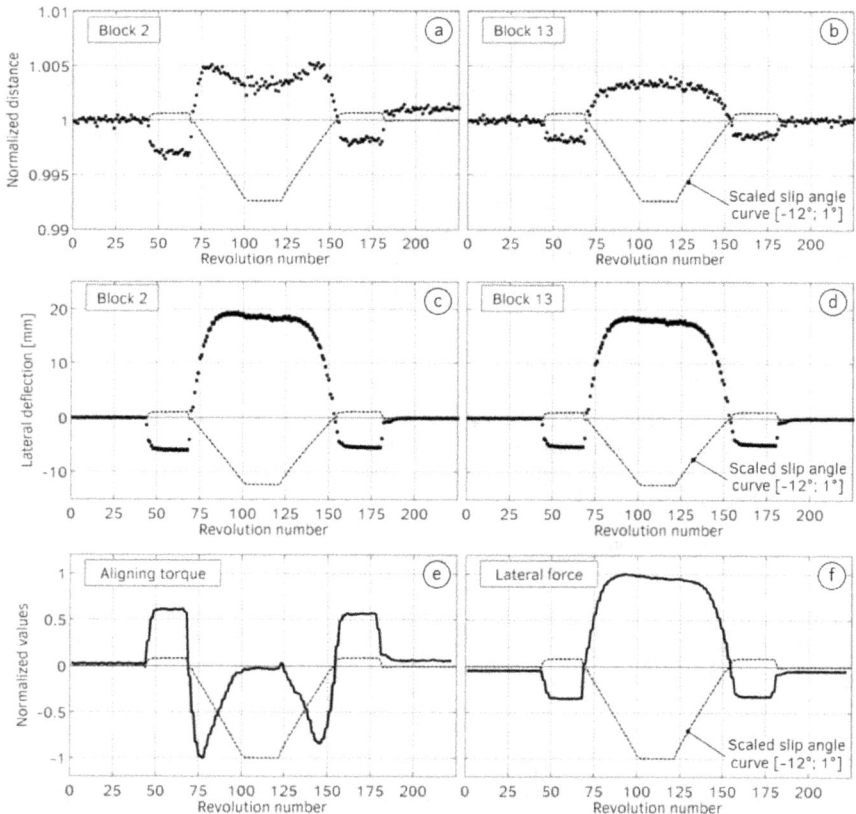

Figure 2.40. Analysis of the correlation between tire deformation and force generation.

The same effects were reproducibly observed for the experiment with a different wheel load, namely 5 kN (Figure 2.41, Appendix A.33-A.34).

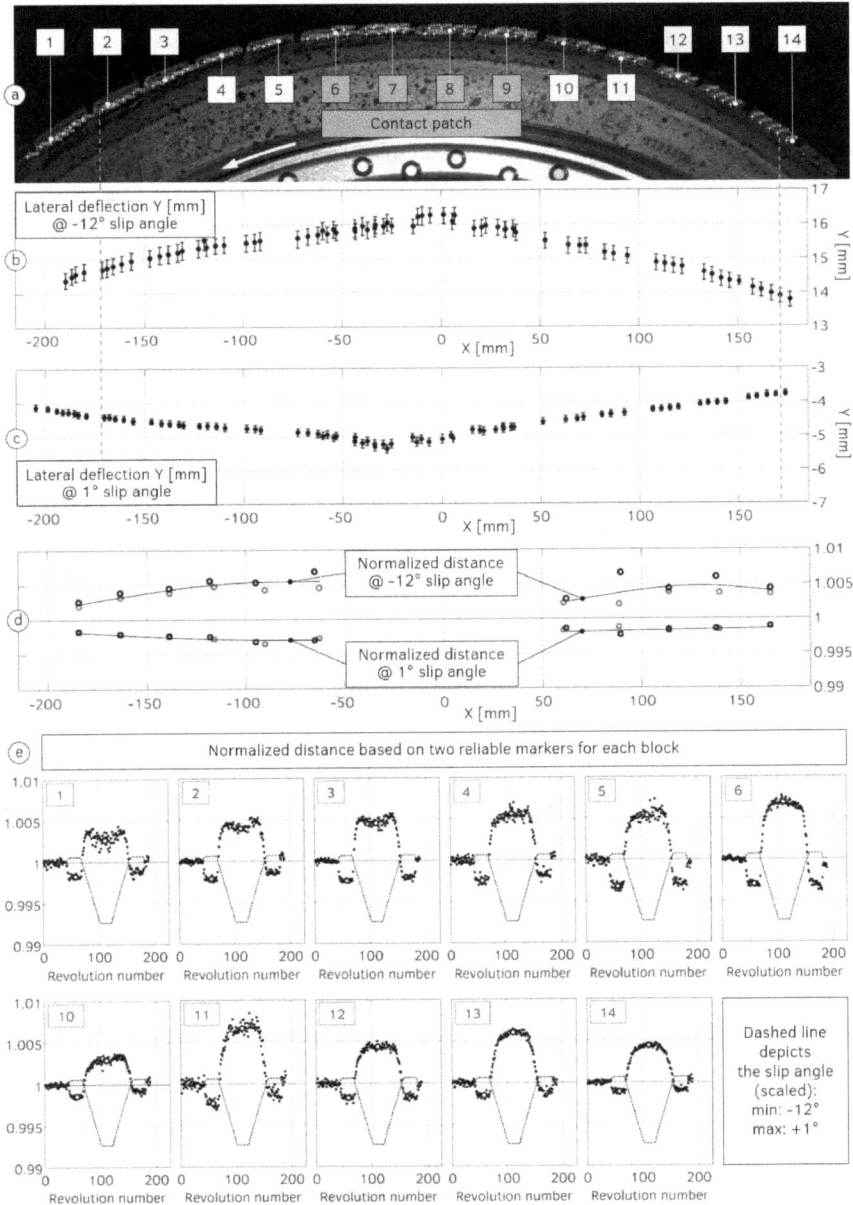

Figure 2.41. Analysis of the carcass deformation behavior during cornering with 5 kN wheel load.

The described method provided the lateral deflection and the change in distance between the cross-sections measured on the carcass shoulder edge. In order to calculate the bending behavior of the carcass body, only one thing was missing – the form of the carcass centerline, which formed a neutral axis of this beam system. These data were gained with help of an in-rim measurement system described in subchapter 2.6.1: Figure 2.42 depicts the lateral deflection observed while cornering, which was measured on the carcass shoulder edge and the centerline.

Due to the limitations of the measuring method and equipment, the centerline deflection was measured at a rolling speed of 24 km/h, and the shoulder edge deflection was captured at a speed of 60 km/h. Therefore, absolute values could not be reasonably compared, but the qualitative deflection shape could be used.

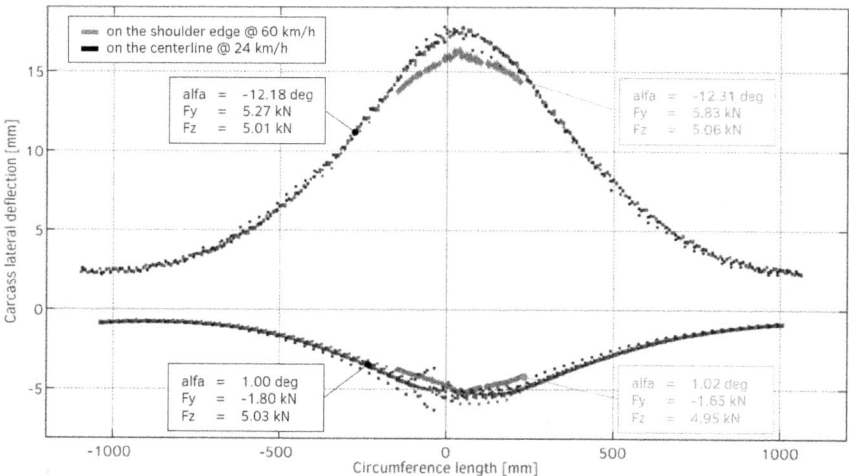

Figure 2.42. Comparison of the lateral deflection of the carcass centerline while cornering at a slip angle of +1° and −12°, measured on the carcass shoulder edge and the centerline.

The curve of the deflected carcass centerline was approximated via a Fourier series. Based on this series, the cross-sections were built according to the shear-free Euler-Bernoulli hypothesis: They remain perpendicular to the neutral axis. Each cross-section formed a straight line from one shoulder edge to another. These are shown in Figure 2.43a (scaled).

Next, the positions of the cross-sections on the shoulder edges were calculated. The change in distance between the cross-sections on the shoulder edge is depicted in Figure 2.43b: This result corresponded to the shear-free bending hypothesis.

In the case of rotation-free assumption, the distance between the cross-sections on the shoulder edge did not change. Therefore, the normalized distances in this case would be equal to 1.

Obtained data made it possible to compare the observed carcass bending behavior with these two hypotheses. The comparison criterion was the normalized distance between the cross-sections on the shoulder edge. Two conditions were considered: Cornering with the slip angle of −12° and with +1°. Figure 2.44 depicts normalized distance that was obtained in different ways:

- Measured optically with the stereocameras;
- Calculated based on deflected centerline of the carcass and shear-free hypothesis;
- Calculated based on deflected centerline of the carcass and shear-free hypothesis and scaled;
- Calculated based on deflected centerline of the carcass and rotation-free hypothesis. This normalized distance corresponds to the horizontal line on the level of 1.

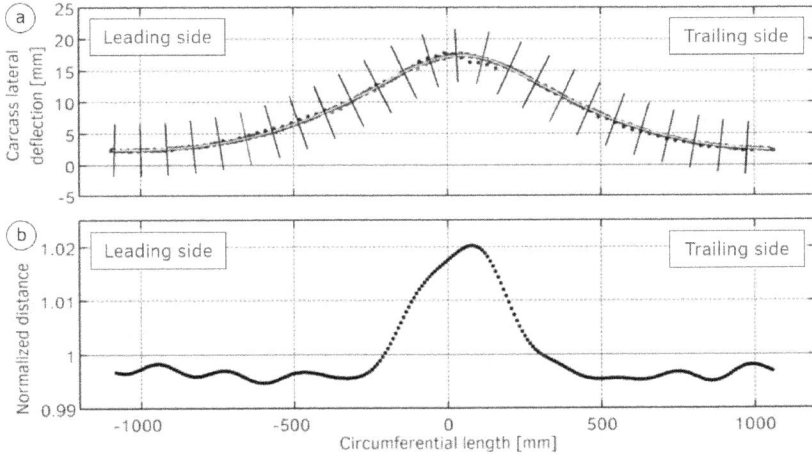

Figure 2.43. Cross-sections of the deflected carcass centerline (a) and the corresponding changes of the normalized distance (b). Cross-sections angle is scaled.

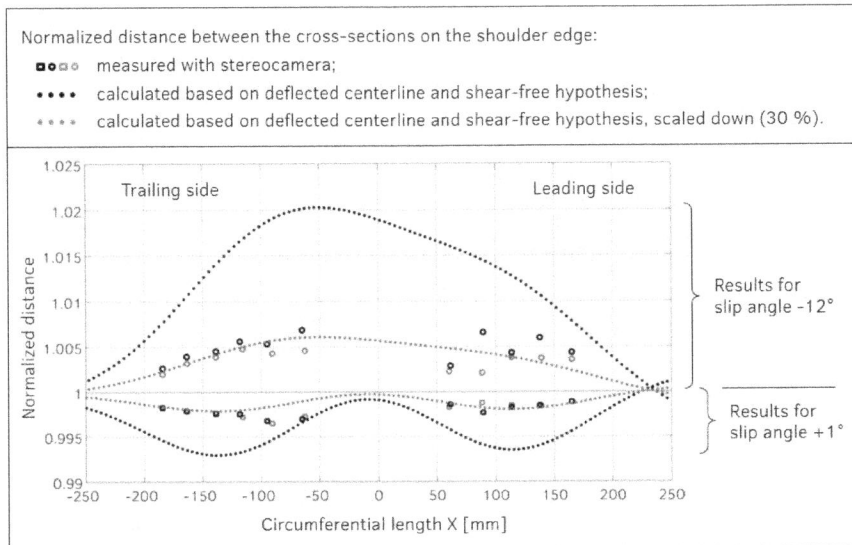

Figure 2.44. Comparison of normalized distance between the cross-sections on the shoulder edge.

According to Figure 2.44, the measured behavior corresponded to the case between these hypotheses, closer to the rotation-free assumption than to the shear-free hypothesis. Therefore, the bending behavior featured both the rotation of the cross-sections and the shear between them (Figure 2.45). The observed orientation angle of the cross sections accounted for 30 % of the orientation angle according to the Euler-Bernoulli hypothesis (shear-free bending). The rest 70% corresponded to the shear angle. This behavior complied with the Timoshenko bending theory.

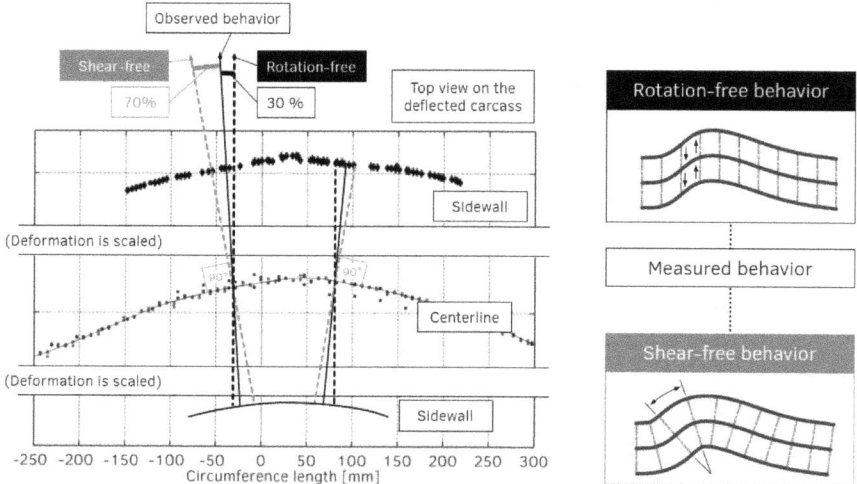

Figure 2.45. Illustration of the observed bending behavior compared with two hypotheses.

At this point, the limitations of the described methodology should be noted. The first issue was the difference of lateral deflection, measured on the shoulder edge with the stereocameras, and the lateral deflection, measured on the carcass centerline with the in-rim cameras. The difference between these achieved 0.7 mm (15 %) for a slip angle of 1° and 1.3 mm (8 %) for a slip angle of −12°. However, the values of the slip angles and the wheel load did correspond well (with a deviation of 2 %). The values of the lateral force differed by 10 %, but not correspondingly to the difference of the wheel load, slip angle or lateral deflection. One significant parameter that differed in these measurements due to method limitations was rolling speed: The deflection of the centerline was measured at 24 km/h, the deflection of the shoulder edge – at 60 km/h. As the normalized distance depends primarily upon the lateral deflection (Figures 2.39-2.41), the influence of the rolling speed on the normalized distance was not significant, but reduced the accuracy of the analysis.

The second issue is connected with an alternative way to investigate bending behavior. Due to the high rigidity of the carcass belt, it could be assumed that the lateral deflection of the carcass centerline and the shoulder edge would be similar. Based on this assumption, the observed orientation angle accounted for 18 % of that according to the shear-free hypothesis. This approach, however, was less accurate. The deflected curve of the carcass centerline was known in this case only for a 350-mm-long area, whereas the in-tire camera system provided these data for the entire 2-m-long perimeter of the carcass. This advantage made it possible to approximate the longer curve and consequently achieve higher precision of its sensitive derivation, which defined the orientation angle of the cross-sections.

In such a manner, the set questions of subchapter 2.6.3 were clarified:

How does the carcass centerline deform in the lateral direction?
- The deflected forms of the carcass centerline were obtained for several use cases (lateral, torsional excitation, cornering). This is important for model development and validation.
- These forms cannot be approximated by curves of a second order and require more complicated description.

What are the numerical values of bending stiffness for the lateral direction?
- Approximation of the deflected carcass with a simplified beam delivered too wide range of the bending stiffness EI. An enhanced beam must be used (with distributed load, tensile force).

How does the wide cylindrical body of the tire carcass deform?
- The carcass bending behavior features both the shear angle and the rotation angle with a ratio of 70:30, respectively, compared to the Euler-Bernoulli hypothesis.

The next keypoint, which is important for understanding and reproducing of the tire force and torque generation, is the physics of the tire tread.

2.7 Tread block properties

When considering the handling properties of a tire, the tread is responsible for two important issues – friction and force generation due to shear. The tribological issues involved here are outside the scope of this research. The shear modulus and Young's modulus of tire tread rubber are dependent of core and surface temperature, excitation frequency and age [Kel12, Soc05]. Detailed material properties are also outside the scope of the current research. To perform an analysis relating to the goals of this investigation, it is necessary to know at least the shear modulus of the tread of the tire sample in the low frequency range. In order to determine this, a shear test setup on the single-axial hydraulic pulsing machine was developed (Figure 2.46, Appendix A.1).

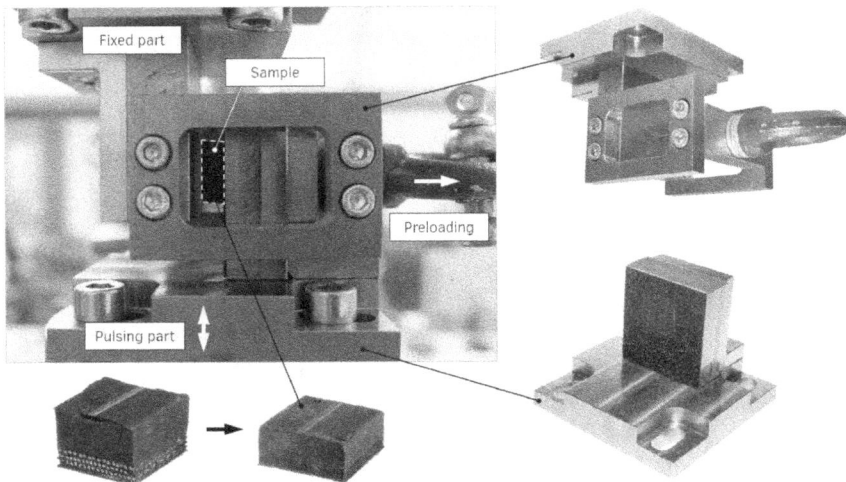

Figure 2.46. Test setup and samples for shearing investigation.

The samples were cut from the preconditioned tire as a block with an area of 20×20 mm.

Analysis of the shear properties was performed using sinus sweep excitation of the sample from 1 to 20 Hz for three values of preloading pressure (0.30, 0.45, 0.60 MPa, Figure 2.47). In this range for the given tire:

- Shear properties did not change significantly (below 10 %);
- Complex shear modulus accounted for 0.67 MPa;
- Storage shear modulus was equal to 0.65 MPa;
- Loss shear modulus was approximately 0.16 MPa.

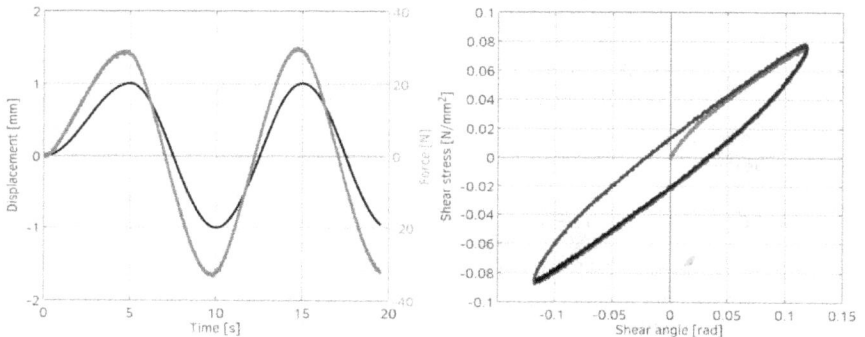

Figure 2.47. Displacement-time diagram, force-time diagram and a hysteresis loop of the tread block shear.

The performed analysis provided nothing about the behavior of the block at the higher frequency range, with different temperatures and on different surfaces. The obtained parameters should therefore be used as a basis for model parameterization, but not as end values. This measurement concludes experimental analysis.

2.8 Summary of chapter 2

The required experimental methods were developed. The necessary observations of the research object were made regarding the following properties:

- tire contact pressure distribution (subchapter 2.4, brief analysis);
- contact patch shape, its change with rolling parameters (subchapter 2.5, detailed analysis);
- lateral flexibility and bending behavior of the carcass (subchapter 2.6, detailed analysis);
- shear properties of the tread (subchapter 2.7, brief analysis).

The observations were interpreted considering limitations of corresponding experimental methods.

The current summary includes only those observations, insights and measured data that are useful for the development of a simulation tool for transient handling analysis.

As the results obtained in different subchapters are interconnected, they are rearranged according to the tire properties and systematized in Table 2.4.

Table 2.4. Summary of the experimental work.

Observation, insight or measured data (what was found out)	Application (what the observation is useful for)
Measured characteristics of the tire: lateral, torsional, cornering stiffness, transient response to slip angle excitation	Model parameterization and validation
The change of the contact patch shape depending upon the rolling speed can be neglected	Model development
Contact patch shape changes significantly depending upon slip angle, it must be considered	Model development
Measured shapes of the contact patch (straight run, cornering)	Model parameterization
Measured deflected forms of the tire carcass: under lateral and torsional excitation, while cornering	Model development and validation
Estimation of the carcass bending stiffness EI in lateral direction cannot be reliably performed with a simplified beam (concentrated forces) and requires more detailed beam description (consideration of distributed load and tensile force)	Model development and parameterization
The distributed lateral stiffness of the carcass can be considered as a displacement- and wheel-load-independent value (0.3 N/mm^2)	Model development and parameterization
With higher wheel load, the lateral stiffness of the tread layer increases proportional to the effective area of the contact patch (15 % per 1 kN)	Model development
The carcass bending features both shear angle and rotation angle with the ratio of 70 % to 30 %, respectively, compared to the Euler-Bernoulli hypothesis (shear-free bending)	Model development and parameterization
Complex shear modulus accounts for 0.67 MPa; storage shear modulus is equal to 0.65 MPa; loss shear modulus is 0.16 MPa.	Model parameterization

Obtained results mean the completion of tasks 1.1 and 1.2 of the mission statement (p. 13). These insights are necessary for the efficient development of the physical model. This tool has to take into account only the physical effects and properties, which are relevant for handling, while remaining simple enough to operate with acceptable calculation time. This challenge is the topic of the next chapter.

3 Simulation method of tire deformation behavior

3.1 Concept development

In chapter 1, simple physical modelling approach was selected to solve the detected scientific problem. Considering the features of the problem and the existing physical models, following requirement was set: The model must be understanding-oriented, based on the physically justified sub-models of carcass and tread. Further analysis of the existing models revealed the feature to be changed to meet this requirement: Consideration of a wide carcass body with improved representation of its bending behavior. Next, two secondary physical effects were identified, which were investigated in detail in chapter 2. As a result, the experimental part of this research provided important insights for model development, namely a displacement- and wheel-load-independent parameter of lateral carcass stiffness, connection between the contact patch shape and the rolling parameters, carcass bending behavior.

Using these insights, a special physical model was developed, although many of its features are shared with existing models. The concept remains a combination of separated carcass and tread sub-models. Their combination is arranged in the following way:

1. The rim is considered as a rigid body.

2. An auxiliary cylindrical surface is constructed coaxially with the rim. It is truncated in the contact patch with a plane parallel to the road surface (Figure 3.1a).

3. The carcass is represented as a flexible belt, which belongs to the received auxiliary surface. The belt is able to slide along this surface only in the axial direction and cannot leave the surface (Figure 3.1a).

4. The belt has a finite non-zero bending stiffness around a radial axis of the wheel. It is preloaded with a constant tensile force along its circumference.

5. The belt connects with the rigid rim in the lateral (axial) direction through an elastic linkage (Figure 3.1b). Such an elastic foundation can be represented with an endless number of independent elastic springs, distributed along the rim circumference (Winkler's elastic foundation, [Wit11, Wun03, Feo99]).

6. The developed mechanical analog is a sub-model of the tire carcass. Such mechanical structure can be called a "wide beam on an elastic foundation" (Figure 3.1c).

7. Brush elements (bristles, [Fro41]) are constructed on the received belt. Each element in an unloaded state is oriented in the radial direction of the wheel (Figure 3.1d). The elements in the contact patch touch the road surface. An element can deflect in a longitudinal and lateral direction without detaching from the road surface (simulating tread shear), unless it achieves the trailing edge.

8. A kinematic constraint between the brush element and the road surface is defined by the adopted friction model. Within the slip limit the element sticks to the road. As soon as the shear force reaches the slip limit, the element slides, but remains in contact with the road.

9. The received mechanical system represents a basic model of a complete tire. The top view (Figure 3.2) depicts two important features: The carcass is not rigid but flexible (unlike the BRIT model [Gip97, Amm97]); the deflected carcass represents a smooth curve without kinks (unlike string-based models [Ell69, Pac12]).

10. Finally, this structure is expanded into a three-dimensional form, in order to take into account tire width (Figure 3.1e-f). The carcass is described not by several independent bodies, but by one coherent belt (unlike Treadsim [Pac12]). The obtained structure represents an enhanced model of a complete tire.

Figure 3.1. Structural scheme of the physical model.

The proposed physical structure meets the requirements identified during the analysis of existing models and those generated by experimental observations. Consequently, it is more detailed. This leads to the increased complexity of both the physical and mathematical representations, which are the subject of the next subchapters.

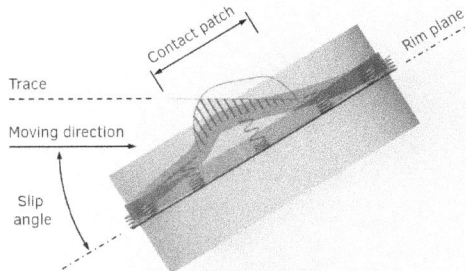

Figure 3.2. Top view on the basic physical model during steady-state cornering.

3.2 Physical representation of the model

The developed model consists of two subsystems:

- The beam on an elastic foundation (carcass model), which carries shear forces from the brush elements;
- The array of the brush elements (tread model), the deformation of which is defined by rim motion law and carcass deflection.

Within each moment in time, these two subsystems are in equilibrium. The logical connection between these subsystems is shown in Figure 3.3, which depicts an order of calculation in a total vehicle simulation.

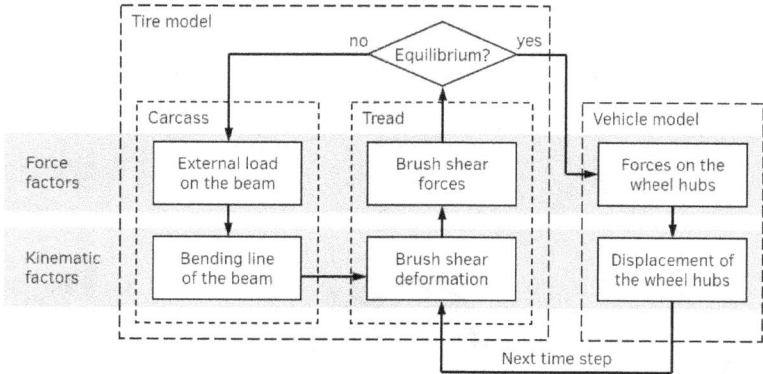

Figure 3.3. Total vehicle simulation scheme.

A typical approach to describe a physical tire model that contains a closed circular body (string, beam), is to figuratively "cut" the tire at its top point and to "unfold" it onto the road surface, as described in Figure 2.26, p. 36. When the developed mechanism is transformed into a flat beam on an elastic foundation, the lateral deflection of the carcass can be described using the beam theory.

In Figure 3.4 an excited physical model of the tire is depicted: The carcass is represented as a flexible beam (1). All forces that act on the beam are transmitted to the one-dimensional neutral axis (2) of this beam. Tire lateral flexibility is described as a distributed elastic connection (3) between the beam neutral axis (2) and the rim plane (4). Tread shear forces are applied on the neutral axis in the form of distributed lateral force (5) and distributed bending torque (6). This torque acts around the axis that is normal to the beam plane (radial axis of the wheel in a given point).

In quasi-static rolling, tread shear deformation develops from the leading edge (A) to the slip-starting point S. The tread is sliding from the point S to the trailing edge (B). Hence, in quasi-static rolling, any part of the carcass body experiences one of three possible load cases (Table 3.1).

Equations of equilibrium are used to determine deformed conditions of flexible bodies in the model at any time step. In the carcass subsystem, this role overtakes the differential equation of a bended beam.

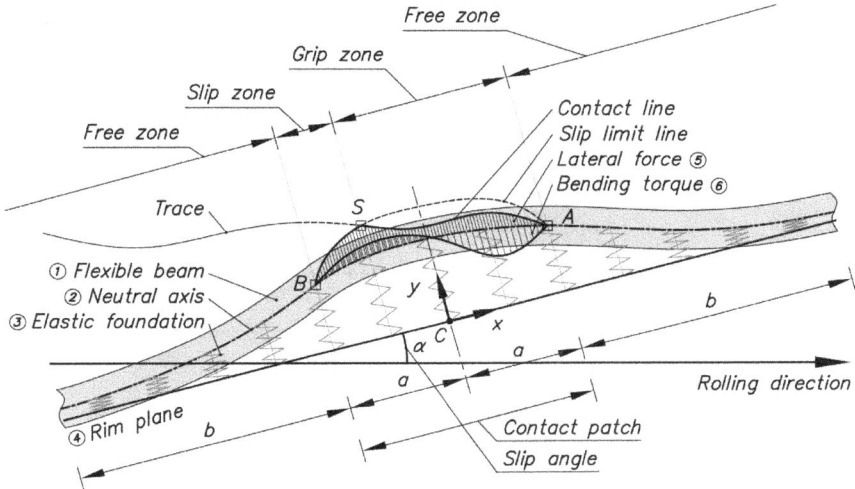

Figure 3.4. Scheme of the tire model when quasi-static cornering without longitudinal slip (unfolded on a flat road surface).

Table 3.1. Three load cases of tire carcass.

Zone	From:	To:	External load on the beam (from the brush elements)
Free zone BA'	Trailing point B	Leading point A' of the next revolution	Absent
Grip zone AS	Leading point A	Slip-starting point S	Defined by the contact line and deflected beam
Slip zone SB	Slip-starting point S	Trailing point B	Defined by the slip limit

The differential equations of beam equilibrium are [Fal69]:

$$\begin{cases} \dfrac{d\vec{Q}}{ds} + \vec{q} = 0 \\[2mm] \dfrac{d\vec{M}}{ds} + \vec{\mu} + \vec{e_1} \times \vec{Q} = 0 \end{cases} \tag{3.1}$$

where:

$\vec{Q}$ and $\vec{M}$ – vectors of internal concentrated force and torque in a cross-section, respectively;

ds – elementary length of the beam: Due to the small deformations, $ds \approx dx$;

$\vec{q}$ and $\vec{\mu}$ – vectors of external distributed force and torque, respectively;

$\vec{e_1}$ – basis vector, oriented along the beam axis.

These vector equations can be resolved into relevant projections in order to separate longitudinal factors from lateral (Figures 3.5-3.6, Equation 3.2).

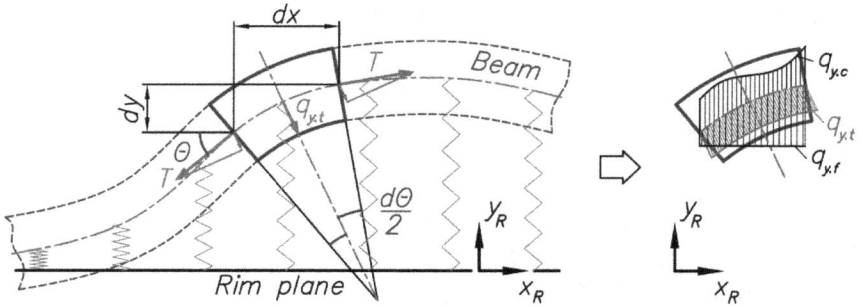

Figure 3.5. Influence of a tensile force on lateral deflection (top view on the carcass).

$$\begin{cases} -\dfrac{dQ_x}{dx} + q_x = 0 \\ -\dfrac{dQ_y}{dx} + q_y = 0 \\ -\dfrac{dM_z}{dx} + \mu_z + Q_y = 0 \end{cases} \qquad (3.2)$$

Lateral deflection of the carcass is essential to describe tire cornering behavior. Therefore, the first equation of (3.2) will not be considered.

For the case of pure bending with small deflections, the Euler-Bernoulli hypothesis can be applied as a reasonable simplification. It connects the curvature of the deflected beam with the bending torque:

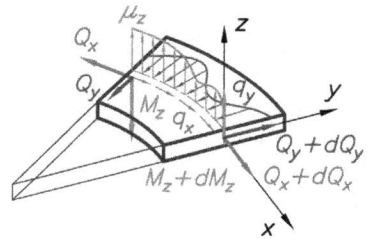

Figure 3.6. Elementary fragment of the flat beam with forces and torques.

$$y'' = \frac{M_z}{EI} \qquad (3.3)$$

where:

$y = y(x)$ – lateral deflection of the beam;

Derivatives are considered with respect to the longitudinal coordinate: $y^{(n)} = \dfrac{d^n y}{dx^n}$;

M_z – concentrated bending torque in the given cross-section;

EI – bending stiffness of the beam in the given cross-section.

Combining equation (3.3) with system (3.2), and assuming that the bending stiffness of the carcass is constant for its whole length, the following equations can be derived:

$$\theta = y'; \qquad M_z = EIy''; \qquad Q_y + \mu_z = EIy'''; \qquad q_y + \mu_z' = EIy^{IV} \qquad (3.4)$$

where:

θ – cross-section rotation angle (Figure 3.5);

M_z and Q_y – concentrated bending torque and lateral force in given cross-section, respectively;

μ_z and q_y – distributed bending torque and lateral force, respectively.

Derivatives are considered with respect to the longitudinal coordinate x.

The final equation of system (3.4) is called the differential equation of a beam bending line. The tire carcass carries several different distributed forces. Hence, the equation must be rewritten as follows:

$$EIy^{IV} = \sum_{i=1}^{n} q_{yi} + \mu_z' \qquad (3.5)$$

where q_{yi} is an i-th distributed lateral force acting on the beam.

Figure 3.5 shows a simplified top view of the deflected carcass and the distributed forces acting on it.

Firstly, the elementary fragment of the carcass is loaded by a flexible force of the elastic foundation $q_{y.f}(x)$. It is proportional to its lateral deflection with a stiffness coefficient k:

$$q_{y.f}(x) = -ky(x) \qquad (3.6)$$

Secondly, the carcass is loaded by tread shear force $q_{y.c}(x)$ in the contact patch. It depends linearly on tread shear deflection (if shear force is lower than static friction limit), or is equal to sliding friction force (if friction limit is achieved).

Finally, the concentrated longitudinal tensile force T in the carcass also influences its lateral deflection (Figure 3.5). This force, for a single elementary fragment of the beam, can be recalculated into lateral force, which should be distributed along the elementary length ds:

$$q_{y.t}(x) = 2T sin\left(\frac{d\theta}{2}\right)\frac{1}{ds} \qquad \rightarrow \qquad q_{y.t}(x) = Ty''(x) \qquad (3.7)$$

where $sin\left(\frac{d\theta}{2}\right) \approx \frac{d\theta}{2}$ and $ds \approx dx$ due to small deflections.

Hence, the differential equation of the beam bending line (3.5) takes the following form:

$$EIy^{IV}(x) - Ty''(x) + ky(x) = q_{y.c}(x) + \mu_z'(x) \qquad (3.8)$$

The derived equation determines the carcass deflection function $y(x)$. Hence, the deformation condition of the first subsystem is determined.

The description of the second subsystem (brush model of the tread) depends upon the carcass deflection function. Each brush element is constructed on the flexible beam of the carcass (root point, Figure 3.7). The other end of the brush element touches the road surface (contact point). Element deformation is therefore the difference between the root and the contact points in horizontal direction.

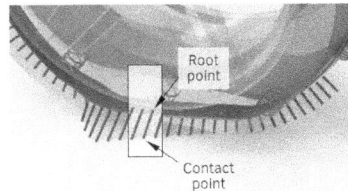

Figure 3.7. Root and contact points.

The positions of the root points are defined by the lateral deflection of the carcass centerline (neutral axis) and the hypothesis regarding the bending behavior of its body (Figure 3.8). The carcass bending behavior is considered with help of a shear angle coefficient k_{ps}, which describes a relation between neutral axis orientation in a point and cross-section orientation in the same point. This term will be considered in the following subchapters.

The positions of the contact points of the elements are defined by kinematic constraints. Initially, these are defined by the first contact of the elements – on the leading edge. As long as the slip limit is not achieved, these points are fixed on the road surface. During rolling of the tire, such a point describes a curve in the coordinate system of the rim (contact line, Figure 3.4). In case of steady-state cornering, this line is straight, inclined towards the rim longitudinal axis with the slip angle (Figure 3.9). When the slip limit is achieved, the contact point is dragged along the road surface. Its position is then defined by its distance from the root point and orientation of a sliding velocity vector.

The area of the contact patch, which changes depending upon the rolling conditions, can be circumscribed by a rectangle $2a \times B$, where a is half of the length of the largest contact patch, and B is width of the largest contact patch.

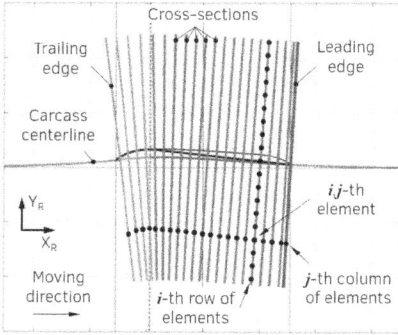

Figure 3.8. Positioning of the root points of the brush elements.

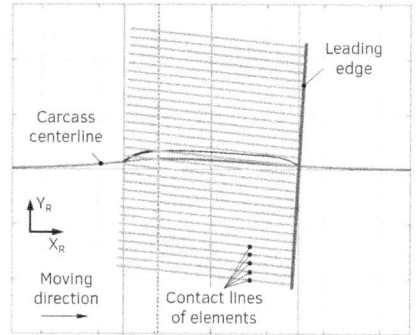

Figure 3.9. Positioning of the contact points of the brush elements (in the grip region).

In the model, this area is discretized to a field of $N_r \times N_e$ elementary fragments, each of which is described via one brush element (Figure 3.10). There are N_r elements in a column and N_e in a row. Thus, element length and width are respectively $l_x = 2a/(N_r - 1)$ and $l_z = B/N_e$. The difference in these expressions has the following reason: Because of the rolling process, the length of the contact patch is counted between centers of the areas of the first and the last elements. The width, on the contrary, is counted between outer borders of the areas of the left-most and the right-most element.

The shear properties of the tread are considered with help of its shear modulus G_t (which is assumed to be constant) and tread height h_t:

$$c_{shear} = \frac{F_{shear}}{\Delta x_{shear}} = \frac{G_t A_{shear}}{h_t}; \qquad (3.9)$$

$$c_s = \frac{G_t(1 \cdot 1)}{h_t}; \qquad c_{xy} = \frac{G_t l_x l_z}{h_t}; \qquad (3.10)$$

where:

c_{shear} – shear stiffness of a block with the area A_{shear}, the height h_t and the shear modulus G_t. Under the force F_{shear} it deflects for Δx_{shear}.

c_s – distributed along the area shear stiffness;

c_{xy} – shear stiffness of the fragment $l_x \times l_z$.

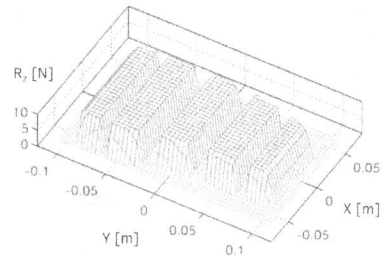

Figure 3.10. Vertical load on the elements.

In such a manner, deflection of a brush element is linearly connected with shear force, which is limited by friction properties. Here, the vertical load on the element is essential. Thus, tire pressure distribution in the contact patch is recalculated to the vertical load on each element $R_z(i,j)$ (Figure 3.10). This can vary depending upon the rolling conditions: By setting $R_z(i,j) = 0$, the shape of the contact patch can be changed if needed.

The summary of the required parameters for the model, input space and output space of its signals is depicted in Figure 3.11.

Input space			Tire model	Output space		
$x_o(t)$ $y_o(t)$ $z_o(t)$	Motion of the wheel center			$F_x(t)$ $F_y(t)$	Longitudinal and lateral forces	
$\gamma(t)$	Wheel camber angle			$M_z(t)$	Aligning torque	
$\phi(t)$	Wheel eigenrotation angle			aux	Carcass deflection	
$\varphi(t)$	Wheel yaw rotation angle			aux	Strain figure, sliding velocities in the contact patch	

Physical model parameters		Geometric model parameters	
EI	Carcass lateral bending stiffness	r	Radius of the free tire
T	Longitudinal tensile force in the carcass	h_t	Height of the tread layer
k_{ps}	Carcass shear angle coefficient		
k	Carcass lateral elasticity		
G_t	Shear modulus of tread material	$2a, B$	Length and width of the contact patch
$R_z(x,y)$	Load distribution in the contact patch	—	Footprint picture

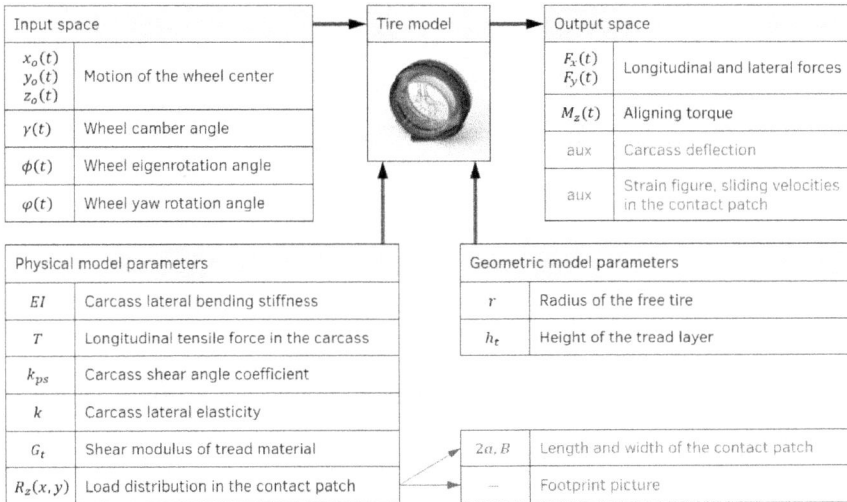

Figure 3.11. Parameter spaces of the model.

Next, the introduced physical mechanism must be represented as a mathematical model. For the described system of the brush elements, equations of kinematic constraint were derived. Together with (3.8), these represent a system of equations that determines the deformation condition of the whole model (beam and brush) for any moment in time. These equations have several properties that exclude the possibility of an analytical solution to this system:

1. The beam deflection becomes a function of two arguments $y(x,t)$. Consequently, its derivations become partial, e.g. $\theta(x,t) = \partial y(x,t)/\partial x$. Hence, the system receives several nonhomogeneous partial differential equations with the maximum order of four.

2. The mathematical consideration of the static friction limit is implemented with a Heaviside step function. Its argument is a ratio of current deflection of a brush element to its maximum deflection, which is an unknown function. Such a Heaviside function makes it impossible to solve the equations by operational calculus methods: It requires to know a Laplace-transform of the sought-for function.

3. An alternative description of the static friction limit is possible with continuous functions, which have close-to-step appearance. However, this makes the system much more sophisticated due to the high order of its argument.

Because of the mentioned issues:

! No reasonable possibility exists to analytically solve the system of the model equations. There is a very limited possibility to solve it using ordinary numerical methods.

! A specific mathematical tool must be developed to numerically solve this system with the required precision.

The following subchapter is devoted to this challenge.

3.3 Model computing method

In order to discretize the continuous-time physical system, it was necessary to analyze the logical sequence of the processes.

1. **In terms of kinematics:** Model input (wheel motion, Figure 3.11) determines displacement of the rigid rim with respect to road surface.

2. This displacement defines tire deformation (which is the sum of carcass deflection and tread shear) and sliding. Hence, kinematic equations constrain whole tire deformation, but the ratio between carcass deflection and tread shear is still unknown.

3. **In terms of dynamics:** The carcass forces, caused by its deflection, and tread forces, caused by its shear, are in equilibrium (no mass is considered). This fact arranges the force-balance equation.

   a. Deflection of the carcass is described by the model "beam on an elastic foundation". Tread shear generates external load on the carcass, which causes carcass deflection.

   b. This external load is the sum of the tread forces and moments of elements from the brush model.

   c. These are determined with help of function of carcass deflection (a) and equations of kinematic constraints with the road surface.

In such a manner, the system of equations of the tire model encloses a closed loop (Figure 3.3). The equations of the subsystem "carcass" and the subsystem "tread" will be considered as separated tasks, assuming the input functions to given subsystem are known.

3.3.1 Computing method of carcass equations

This subchapter is devoted to identifying the deflection of the carcass, described as a beam system. Incoming data are the lateral and longitudinal forces in each brush element. Outcoming result is the continuous function of the carcass deflection, which results from the external load (incoming data).

The described task means solving equation (3.8). The principal possibility of the analytic solution depends upon the appearance of the function in the right-hand side of equation (3.8): It is the sum of the lateral distributed load $q_{y,c}(x)$ and the derivative of the distributed bending torque $\mu_z{}'(x)$ with respect to the longitudinal coordinate x. First, the integrability of this function must be checked:

1. **The necessary condition:** The function must be bounded on a compact interval.

   a. The interval is limited by the length of the contact patch. Hence, it is compact.

   b. The function of the external load $q_{y,c}(x) + \mu_z{}'(x)$ is limited by friction properties. Hence, the function is bounded.

2. **The sufficient condition:** The bounded function in a compact interval must be continuous at almost every point (a number of discontinuity points must be finite).

   a. Both summands, $q_{y,c}(x)$ and $\mu_z{}'(x)$, are piecewise continuous functions.

Hence, the function is integrable within the interval of the contact patch.

Considering the flat system "beam on an elastic foundation", it was reasonable to separate the whole beam into two parts: The contact part (within the contact patch) and the free part (outside the contact patch, Figure 3.12). As a matter of convenience, the length of the free part was designated $2b$, length of the contact part – $2a$. The deflected carcass was described in the coordinate system Oxy. Axis Ox coincided with axis CX_C (wheel longitudinal axis, Figure 1.15).

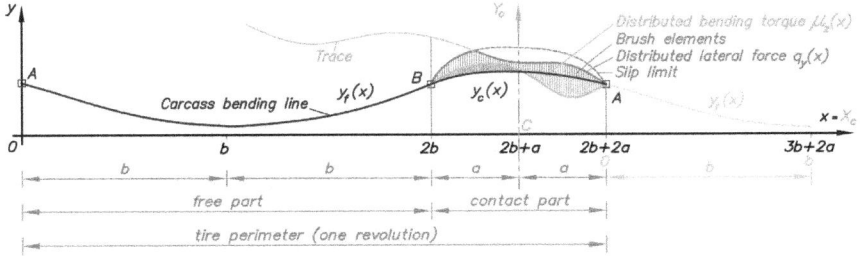

Figure 3.12. Scheme of the carcass deformation as a beam on an elastic foundation.

There is no external load on the carcass in the free part of the tire. Therefore, equation (3.8) transforms into an homogeneous form:

$$El\,y_f^{IV}(x) - T y_f''(x) + k y_f(x) = 0 \tag{3.11}$$

where:

$y_f(x)$ – a function of the lateral deflection of the tire carcass in the free part;

x – longitudinal coordinate.

Solution $y_f(x)$ of this equation in the general case consists of the homogeneous part only:

$$y_f(x) = C_{f1}e^{\lambda_{f1}x} + C_{f2}e^{-\lambda_{f1}x} + C_{f3}e^{\lambda_{f3}x} + C_{f4}e^{-\lambda_{f3}x} \tag{3.12}$$

where:

$C_{f1...4}$ – constant coefficients (free part), which can be found using boundary conditions;

$\lambda_{f1,3} = \sqrt{\dfrac{T \pm \sqrt{T^2 - 4Elk}}{2El}}$ – solutions of the characteristic polynomial of the equation in the free part.

In the contact part, the tire is loaded with lateral force and bending torque from the tread. Therefore, equation (3.8) for this part remains inhomogeneous:

$$El\,y_c^{IV}(x) - T y_c''(x) + k y_c(x) = q_{y.c}(x) + \mu_z'(x) \tag{3.13}$$

Hence, its solution $y_c(x)$ receives an additional inhomogeneous part $y_{c.ih}(x)$:

$$y_c(x) = C_{c1}e^{\lambda_{c1}x} + C_{c2}e^{-\lambda_{c1}x} + C_{c3}e^{\lambda_{c3}x} + C_{c4}e^{-\lambda_{c3}x} + y_{c.ih}(x) \tag{3.14}$$

where:

$C_{c1...4}$ – constant coefficients (contact part), which can be found using boundary conditions;

$\lambda_{c1,3} = \sqrt{\dfrac{T \pm \sqrt{T^2 - 4Elk}}{2El}}$ – solutions of the characteristic polynomial of the equation in the contact part.

Boundary conditions originate from the closed structure of the beam: In the connection points between the beam in the contact part and the beam in the free part (A and B, Figure 3.12), these beams must coincide, be tangential, have the same curvature and same curvature change rate:

$$\begin{cases} y_f^{(0...3)}(x)\big|_{x=0} = y_c^{(0...3)}(x)\big|_{x=2b+2a} \\ y_f^{(0...3)}(x)\big|_{x=2b} = y_c^{(0...3)}(x)\big|_{x=2b} \end{cases} \tag{3.15}$$

where:

$x = 0$ and $x = 2b + 2a$ – coordinates of point A due to the closed structure (Figure 3.12);

$x = 2b$ – coordinate of point B.

Introducing equations (3.12) and (3.14) to the boundary conditions (3.15), a system of linear equations can be composed. As a matter of convenience, this system can be separated into two matrix equations – one for first and one for second line of the system (3.15):

$$\begin{cases} M_{fc1} \cdot [C_{f1} \quad C_{f2} \quad C_{f3} \quad C_{f4} \quad C_{c1} \quad C_{c2} \quad C_{c3} \quad C_{c4}]^T = R_{fc1} \\ M_{fc2} \cdot [C_{f1} \quad C_{f2} \quad C_{f3} \quad C_{f4} \quad C_{c1} \quad C_{c2} \quad C_{c3} \quad C_{c4}]^T = R_{fc2} \end{cases} \tag{3.16}$$

The matrix of coefficients M_{fc1} and the vector of the fixed terms R_{fc1} for the first four boundary conditions (3.15):

$$M_{fc1} = \begin{bmatrix} +\lambda_{f1}^0 & +\lambda_{f1}^0 & +\lambda_{f3}^0 & +\lambda_{f3}^0 & \dots \\ +\lambda_{f1}^1 & -\lambda_{f1}^1 & +\lambda_{f3}^1 & -\lambda_{f3}^1 & \dots \\ +\lambda_{f1}^2 & +\lambda_{f1}^2 & +\lambda_{f3}^2 & +\lambda_{f3}^2 & \dots \\ +\lambda_{f1}^3 & -\lambda_{f1}^3 & +\lambda_{f3}^3 & -\lambda_{f3}^3 & \dots \end{bmatrix}$$

$$\begin{matrix} \dots & -\lambda_{c1}^0 e^{\lambda_{c1}(2a+2b)} & -\lambda_{c1}^0 e^{-\lambda_{c1}(2a+2b)} & -\lambda_{c3}^0 e^{\lambda_{c3}(2a+2b)} & -\lambda_{c3}^0 e^{-\lambda_{c3}(2a+2b)} \\ \dots & -\lambda_{c1}^1 e^{\lambda_{c1}(2a+2b)} & +\lambda_{c1}^1 e^{-\lambda_{c1}(2a+2b)} & -\lambda_{c3}^1 e^{\lambda_{c3}(2a+2b)} & +\lambda_{c3}^1 e^{-\lambda_{c3}(2a+2b)} \\ \dots & -\lambda_{c1}^2 e^{\lambda_{c1}(2a+2b)} & -\lambda_{c1}^2 e^{-\lambda_{c1}(2a+2b)} & -\lambda_{c3}^2 e^{\lambda_{c3}(2a+2b)} & -\lambda_{c3}^2 e^{-\lambda_{c3}(2a+2b)} \\ \dots & -\lambda_{c1}^3 e^{\lambda_{c1}(2a+2b)} & +\lambda_{c1}^3 e^{-\lambda_{c1}(2a+2b)} & -\lambda_{c3}^3 e^{\lambda_{c3}(2a+2b)} & +\lambda_{c3}^3 e^{-\lambda_{c3}(2a+2b)} \end{matrix} \tag{3.17}$$

$$R_{fc1} = \begin{bmatrix} 1 & \dfrac{d}{dx} & \dfrac{d^2}{dx^2} & \dfrac{d^3}{dx^3} \end{bmatrix}^T \cdot y_{c.ih}(x)\big|_{x=2a+2b} \tag{3.18}$$

The matrix of coefficients M_{fc2} and the vector of constant terms R_{fc2} for the last four boundary conditions (3.15):

$$M_{fc2} = \begin{bmatrix} +\lambda_{f1}^0 e^{\lambda_{f1}2b} & +\lambda_{f1}^0 e^{-\lambda_{f1}2b} & +\lambda_{f3}^0 e^{\lambda_{f3}2b} & +\lambda_{f3}^0 e^{-\lambda_{f3}2b} & \dots \\ +\lambda_{f1}^1 e^{\lambda_{f1}2b} & -\lambda_{f1}^1 e^{-\lambda_{f1}2b} & +\lambda_{f3}^1 e^{\lambda_{f3}2b} & -\lambda_{f3}^1 e^{-\lambda_{f3}2b} & \dots \\ +\lambda_{f1}^2 e^{\lambda_{f1}2b} & +\lambda_{f1}^2 e^{-\lambda_{f1}2b} & +\lambda_{f3}^2 e^{\lambda_{f3}2b} & +\lambda_{f3}^2 e^{-\lambda_{f3}2b} & \dots \\ +\lambda_{f1}^3 e^{\lambda_{f1}2b} & -\lambda_{f1}^3 e^{-\lambda_{f1}2b} & +\lambda_{f3}^3 e^{\lambda_{f3}2b} & -\lambda_{f3}^3 e^{-\lambda_{f3}2b} & \dots \end{bmatrix}$$

$$\begin{matrix} \dots & -\lambda_{c1}^0 e^{\lambda_{c1}2b} & -\lambda_{c1}^0 e^{-\lambda_{c1}2b} & -\lambda_{c3}^0 e^{\lambda_{c3}2b} & -\lambda_{c3}^0 e^{-\lambda_{c3}2b} \\ \dots & -\lambda_{c1}^1 e^{\lambda_{c1}2b} & +\lambda_{c1}^1 e^{-\lambda_{c1}2b} & -\lambda_{c3}^1 e^{\lambda_{c3}2b} & +\lambda_{c3}^1 e^{-\lambda_{c3}2b} \\ \dots & -\lambda_{c1}^2 e^{\lambda_{c1}2b} & -\lambda_{c1}^2 e^{-\lambda_{c1}2b} & -\lambda_{c3}^2 e^{\lambda_{c3}2b} & -\lambda_{c3}^2 e^{-\lambda_{c3}2b} \\ \dots & -\lambda_{c1}^3 e^{\lambda_{c1}2b} & +\lambda_{c1}^3 e^{-\lambda_{c1}2b} & -\lambda_{c3}^3 e^{\lambda_{c3}2b} & +\lambda_{c3}^3 e^{-\lambda_{c3}2b} \end{matrix} \tag{3.19}$$

$$R_{fc2} = \begin{bmatrix} 1 & \dfrac{d}{dx} & \dfrac{d^2}{dx^2} & \dfrac{d^3}{dx^3} \end{bmatrix}^T \cdot y_{c.ih}(x)\big|_{x=2b} \tag{3.20}$$

Thus, the system of linear equations (3.16) has a dimension of 8×8 with eight unknowns. Hence, it is consistent.

The appearance of an inhomogeneous solution $y_{c.ih}(x)$ of equation (3.14) depends upon the right-hand side function of equation (3.13): $q_{y.c}(x) + \mu_z'(x)$. According to the model structure, this function was defined by the tread shear for each brush element in the contact patch. Thus, it was not continuous, but discrete. However, the inhomogeneous solution $y_{c.ih}(x)$ of equation (3.14) was required as a continuous function. In order to analytically identify this inhomogeneous solution, it was necessary to approximate the discrete right-hand side function $q_i = q_{y.c}(x_i) + \mu_z'(x_i)$ with a continuous function, the appearance of which makes it possible to identify the inhomogeneous solution for equation (3.14).

Due to transient excitation of tire, the function $q_i = q_{y.c}(x_i) + \mu_z'(x_i)$ is not monotone. It may have steps as well as linear regions (Figure 3.13). The following methods to solve this task were analyzed regarding the computation time: collocation method, approximation via linear segments, approximation via Fourier series, approximation via Fourier series combined with a polynomial, and approximation with Fast Fourier Transform (FFT). This analysis [Sar16] showed that the fastest option was to approximate the discrete function q_i with one linear segment between each two values of a function (Figure 3.14). Such interval is the distance between two neighboring brush elements.

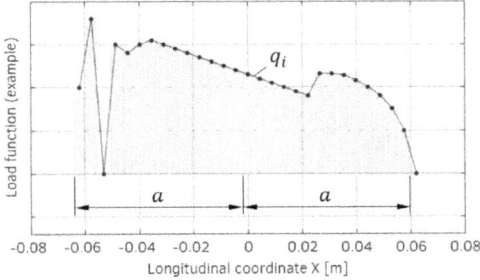

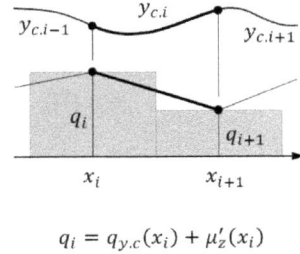

Figure 3.13. An example of the load function q_i.

Figure 3.14. Approximation with linear segments.

$$q_i = q_{y.c}(x_i) + \mu_z'(x_i)$$

According to this approach, the solution of the differential equation for each interval $y_{c.i}(x)$ was:

$$y_{c.i}(x) = C_{c1}e^{\lambda_{c1}x} + C_{c2}e^{-\lambda_{c1}x} + C_{c3}e^{\lambda_{c3}x} + C_{c4}e^{-\lambda_{c3}x} + a_{i1}x + a_{i0} \qquad (3.21)$$

where:

$a_{i1}x + a_{i0} = y_{c.ih.i}(x)$ – inhomogeneous part of solution (different for each interval $[x_i; x_{i+1}]$).
The coefficients for each interval were determined with:

$$\begin{cases} k(a_{1.i}x_i + a_{0.i}) = q_i \\ k(a_{1.i}x_{i+1} + a_{0.i}) = q_{i+1} \end{cases} \qquad (3.22)$$

Solution $y_{c.i}(x)$ represented beam deflection. Hence, at each point between two intervals, the same boundary conditions were valid as conditions (3.15):

$$\forall i = 2,3 \dots (N_r - 1): \qquad y_{c.i-1}^{(0\dots3)}(x)\big|_{x=x_i} = y_{c.i}^{(0\dots3)}(x)\big|_{x=x_i} \qquad (3.23)$$

At the first ($i = 1$) and the last ($i = N_r$) points of the contact patch, these boundary conditions were:

$$i = 1: \qquad y_f^{(0\dots3)}(x)\big|_{x=2b+2a} = y_{c.1}^{(0\dots3)}(x)\big|_{x=2b+2a} \qquad (3.24)$$

$$i = N_r: \qquad y_f^{(0\dots3)}(x)\big|_{x=2b} = y_{c.N_r}^{(0\dots3)}(x)\big|_{x=2b} \qquad (3.25)$$

Due to the linear form of the inhomogeneous part, the boundary conditions (3.23) transformed to the following:

$$\forall i = 2,3 \dots (N_r - 1): \quad \begin{bmatrix} \sum_{j=1}^{4} (C_{c.j.i} - C_{c.j.i-1}) \lambda_{c.j}^0 e^{\lambda_{c.j} x_i} \\ \sum_{j=1}^{4} (C_{c.j.i} - C_{c.j.i-1}) \lambda_{c.j}^1 e^{\lambda_{c.j} x_i} \\ \sum_{j=1}^{4} (C_{c.j.i} - C_{c.j.i-1}) \lambda_{c.j}^2 e^{\lambda_{c.j} x_i} \\ \sum_{j=1}^{4} (C_{c.j.i} - C_{c.j.i-1}) \lambda_{c.j}^3 e^{\lambda_{c.j} x_i} \end{bmatrix} = \begin{bmatrix} 0 \\ 0 \\ 0 \\ 0 \end{bmatrix} \tag{3.26}$$

In the first and the last points of the contact patch, the boundary conditions transformed to:

$$i = 1 \Rightarrow x_i = 2b + 2a: \quad \begin{bmatrix} \sum_{j=1}^{4} C_{c.j.1} \lambda_{c.j}^0 e^{\lambda_{c.j} x_i} \\ \sum_{j=1}^{4} C_{c.j.1} \lambda_{c.j}^1 e^{\lambda_{c.j} x_i} \\ \sum_{j=1}^{4} C_{c.j.1} \lambda_{c.j}^2 e^{\lambda_{c.j} x_i} \\ \sum_{j=1}^{4} C_{c.j.1} \lambda_{c.j}^3 e^{\lambda_{c.j} x_i} \end{bmatrix} - \begin{bmatrix} \sum_{j=1}^{4} C_{f.j} \lambda_{f.j}^0 e^{\lambda_{f.j} x_i} \\ \sum_{j=1}^{4} C_{f.j} \lambda_{f.j}^1 e^{\lambda_{f.j} x_i} \\ \sum_{j=1}^{4} C_{f.j} \lambda_{f.j}^2 e^{\lambda_{f.j} x_i} \\ \sum_{j=1}^{4} C_{f.j} \lambda_{f.j}^3 e^{\lambda_{f.j} x_i} \end{bmatrix} = - \begin{bmatrix} a_{1.1} x_i + a_{0.1} \\ a_{1.1} \\ 0 \\ 0 \end{bmatrix} \tag{3.27}$$

$$i = N_r \Rightarrow x_i = 2b: \quad \begin{bmatrix} \sum_{j=1}^{4} C_{f.j} \lambda_{f.j}^0 e^{\lambda_{f.j} x_i} \\ \sum_{j=1}^{4} C_{f.j} \lambda_{f.j}^1 e^{\lambda_{f.j} x_i} \\ \sum_{j=1}^{4} C_{f.j} \lambda_{f.j}^2 e^{\lambda_{f.j} x_i} \\ \sum_{j=1}^{4} C_{f.j} \lambda_{f.j}^3 e^{\lambda_{f.j} x_i} \end{bmatrix} - \begin{bmatrix} \sum_{j=1}^{4} C_{c.j.N_r} \lambda_{c.j}^0 e^{\lambda_{c.j} x_i} \\ \sum_{j=1}^{4} C_{c.j.N_r} \lambda_{c.j}^1 e^{\lambda_{c.j} x_i} \\ \sum_{j=1}^{4} C_{c.j.N_r} \lambda_{c.j}^2 e^{\lambda_{c.j} x_i} \\ \sum_{j=1}^{4} C_{c.j.N_r} \lambda_{c.j}^3 e^{\lambda_{c.j} x_i} \end{bmatrix} = \begin{bmatrix} a_{1.N_r} x_i + a_{0.N_r} \\ a_{1.N_r} \\ 0 \\ 0 \end{bmatrix} \tag{3.28}$$

The composition of (3.26-3.28) resulted in a system of eight linear equations, which defined eight coefficients $[C_{c.1.1} \ \ C_{c.2.1} \ \ C_{c.3.1} \ \ C_{c.4.1} \ \ C_{f.1} \ \ C_{f.2} \ \ C_{f.3} \ \ C_{f.4}]^T$. Next, using (3.26), the remaining coefficients $C_{c.j.i}$ were defined, where $i = 2,3 \dots N_r$; $j = 1 \dots 4$.

In this way, the solution to the inhomogeneous differential equation (3.13) for each interval were found. The homogeneous part was the same for all intervals. Hence, in order to obtain a continuous function of an inhomogeneous solution, it was necessary to approximate the linear functions of the inhomogeneous part of the solution for each interval. A suitable tool for this purpose was approximation using a Fourier series. The carcass deflection function, which was the solution of (3.13), was represented as the following:

$$y_c(x) = C_{c1} e^{\lambda_{c1} x} + C_{c2} e^{-\lambda_{c1} x} + C_{c3} e^{\lambda_{c3} x} + C_{c4} e^{-\lambda_{c3} x} +$$

$$+ \frac{a_0}{2} + \sum_{n=1}^{n_h} \left(a_n \cos \frac{\pi n (x - 2b - a)}{a} + b_n \sin \frac{\pi n (x - 2b - a)}{a} \right) \tag{3.29}$$

where:

$y_{c.ih}(x) = \frac{a_0}{2} + \sum_{n=1}^{n_h} \left(a_n \cos \frac{\pi n (x-2b-a)}{a} + b_n \sin \frac{\pi n (x-2b-a)}{a} \right)$ – inhomogeneous part of solution as the Fourier series on the interval of contact patch $x = [-a; a] + 2b + a$;

a_n, b_n – Fourier coefficients;

n_h – number of harmonic components.

Fourier coefficients were calculated in the following way:

$$a_n = \sum_{i=1}^{N_r} \left[\frac{1}{a} \int_{x_i}^{x_{i+1}} [a_{1,i} x_i + a_{0,i}] \cos \frac{\pi n (x - 2b - a)}{a} dx \right] \tag{3.30}$$

$$b_n = \sum_{i=1}^{N_r} \left[\frac{1}{a} \int_{x_i}^{x_{i+1}} [a_{1,i} x_i + a_{0,i}] \sin \frac{\pi n (x - 2b - a)}{a} dx \right] \tag{3.31}$$

This step completed calculation of carcass deflection. For any given load on the carcass, the described routine provided carcass deflection as two continuous functions – one for the free part of the tire and one for its contact part. The form of the carcass deflection function was not limited by the form of the approximation function (unlike the parabola in existing models, 1.3). This measure is necessary for gaining a deeper understanding of the processes involved in a rolling tire. Hence, task 2.1.1 (p. 13) was solved. The second part of the model computation routine refers to the tread subsystem.

3.3.2 Computing method of tread equations

This subchapter is devoted to identifying the shear deformation of tread, described as an array of brush elements. Incoming data are a function of carcass deflection and kinematic constraints: positions of the tread contact points in the previous time step (Figure 3.7) and rim motion laws. Outcoming result is lateral and longitudinal forces in each brush element.

As described in Figure 3.8, the contact patch was discretized to an array of $N_r \times N_e$ brush elements. For any given moment of time, the following data were assumed to be known:

$x_{ct0}(i,j), y_{ct0}(i,j)$ – position of the contact point of i,j-th brush element <u>in the previous time step</u>;

$R_z(i,j)$ – vertical load on the i,j-th brush element;

$x(t), y(t), \varphi(t), \gamma(t), \phi(t)$ – rim coordinates including time derivatives $\dot{x}(t), \dot{y}(t), \dot{\varphi}(t), \dot{\gamma}(t), \dot{\phi}(t)$;

$y_c(x)$ – carcass deflection function in the contact patch (Figure 3.15).

The time step was selected so that with each step one row of the brush elements came into the contact patch:

$$dt = \frac{l_x}{\dot{x}(t)} \tag{3.32}$$

As observed in 2.6.3, carcass bending behavior features significant shear angle and does not follow the Euler-Bernoulli hypothesis. This behavior complies with the Timoshenko bending theory. However, its application introduces a second derivative of external load function into the differential equation of bending, which cannot be defined with sufficient precision because of the discreteness of this function. Thus, calculation of the deflection of the carcass neutral axis was performed based on the Euler-Bernoulli beam theory. To determine the orientation of the carcass cross-sections, a simplified approach to consider shear angle was adopted: Using the constant, experimentally identified coefficient k_{ps}, orientation angle $\theta(i)$ of the cross-sections was correspondingly reduced (Figure 3.15a):

$$\theta(i) = k_{ps} \, atan\left[\frac{d}{dx}y_c(x)\right]\Bigg|_{x=2b+2a-(i-1)\cdot l_x} \tag{3.33}$$

where:

$\theta(i)$ – orientation angle of a beam cross-section, located in $x_i = 2b + 2a - (i-1) \cdot l_x$;

$atan\left[\frac{d}{dx}y_c(x)\right]$ – orientation angle of the tangent to the bent neutral axis $y_c(x)$;

$k_{ps} = 0.3$ – coefficient, which considers the presence of the shear angle and reduces the rotation angle corresponding to experimental results (subchapter 2.6.3);

The experimental data showed that this rate (k_{ps}) does not change perceptibly within the contact patch and can be assumed to be constant for low and high values of carcass deflection (Figure 2.43). Hence, this assumption is fair for the set goals of this research.

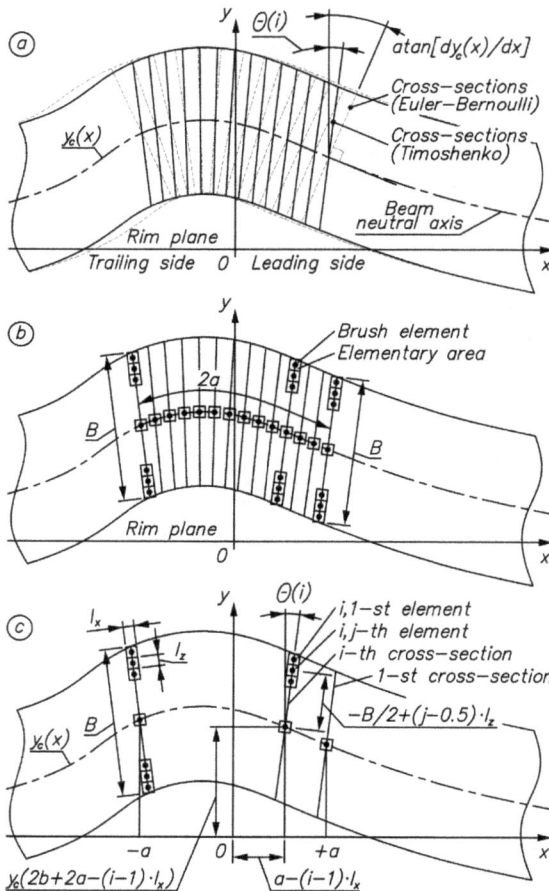

Figure 3.15. Calculation of the coordinates of the brush elements.

The position of the root points of the i,j-th brush element (Figure 3.7) was calculated in the following way (Figure 3.15):

$$x_r(i,j) = a - (i-1) \cdot l_x + \left[-\frac{B}{2} + (j - 0.5) \cdot l_z\right] \sin(\theta(i))$$

$$y_r(i,j) = y_c(2b + 2a - (i-1) \cdot l_x) + \left[-\frac{B}{2} + (j - 0.5) \cdot l_z\right] \cos(\theta(i))$$

(3.34)

where:

B – width of the contact patch (Figure 3.15b).

The positions of the contact points (Figure 3.7) of the brush elements were calculated in the following sequence:

$$\begin{bmatrix} x_{ct}(i,j) \\ y_{ct}(i,j) \end{bmatrix} = \begin{bmatrix} \cos(\dot\varphi(t)dt) & -\sin(\dot\varphi(t)dt) \\ \sin(\dot\varphi(t)dt) & \cos(\dot\varphi(t)dt) \end{bmatrix} \begin{bmatrix} x_{ct0}(i,j) - \dot x(t)dt \\ y_{ct0}(i,j) + \dot y(t)dt \end{bmatrix}$$

(3.35)

where:

$x_{ct}(i,j)$, $y_{ct}(i,j)$ – position of the contact point of the i,j-th brush element in the current time step;

$\dot\varphi(t)dt$ – elementary angle of rim rotation around the vertical axis within the time step, which introduces the rotation matrix;

$\dot x(t)dt$ and $\dot y(t)dt$ – the elementary longitudinal and lateral rim displacement within the time step.

Next, the shear deflection of each brush element was calculated. Firstly, kinematic deflection was defined, assuming the absence of a friction limit:

$$\begin{cases} d_{kx}(i,j) = x_{ct}(i,j) - x_r(i,j) \\ d_{ky}(i,j) = y_{ct}(i,j) - y_r(i,j) \end{cases}$$

(3.36)

where:

$d_{kx}(i,j)$, $d_{ky}(i,j)$ – longitudinal and lateral kinematic deflection of a brush element.

Then, the magnitude $d_k(i,j)$ and orientation angle $\varphi_k(i,j)$ of the deflection vector were determined:

$$d_k(i,j) = \sqrt{\left(d_{kx}(i,j)\right)^2 + \left(d_{ky}(i,j)\right)^2}$$

(3.37)

$$\varphi_k(i,j) = \begin{cases} \text{sign}(d_{kx}(i,j))\dfrac{\pi}{2} & if \quad d_{kx}(i,j) = 0 \\ \text{atan}\left(\dfrac{d_{ky}(i,j)}{d_{kx}(i,j)}\right) & if \quad d_{kx}(i,j) > 0 \\ \pi + \text{atan}\left(\dfrac{d_{ky}(i,j)}{d_{kx}(i,j)}\right) & if \quad d_{kx}(i,j) < 0 \end{cases}$$

(3.38)

Subsequently, the kinematic deflection values were compared with the slip limit in order to calculate the real deflection values $d_x(i,j)$ and $d_y(i,j)$ (real means with consideration of slip):

$$\begin{bmatrix} d_x(i,j) \\ d_y(i,j) \end{bmatrix} = \begin{cases} \begin{bmatrix} d_{kx}(i,j) \\ d_{ky}(i,j) \end{bmatrix} & if \quad d_k(i,j) \leq \mu R_z(i,j) \\ \mu R_z(i,j) \begin{bmatrix} \cos(\varphi_k(i,j)) \\ \sin(\varphi_k(i,j)) \end{bmatrix} & if \quad d_k(i,j) > \mu R_z(i,j) \end{cases}$$

(3.39)

With these values, the distributed lateral load $q_{y.c}(i)$ and distributed bending torque $\mu_z(i)$ were found:

$$i = 1,2 \dots N_r \qquad q_{y.c}(i) = c_{xy} \sum_{j=1}^{N_e} d_y(i,j)$$

(3.40)

$$i = 1,2 \dots N_r \quad \mu_z(i) = c_{xy} \sum_{j=1}^{N_e} \left[\left(-d_x(i,j) \cos\theta(i) - d_y(i,j) \sin\theta(i) \right) \left(-\frac{B}{2} + (j - 0.5) \cdot l_z \right) \right] \quad (3.41)$$

where:

c_{xy} – shear stiffness of the fragment $l_x \times l_z$, Equation (3.10).

Hence, the required outgoing data of the tread subsystem were found. Task 2.1.2 (p. 13) was therefore solved. The only remaining task was to couple two described computing routines into one loop. The solving of this computing loop is a subject of the next subchapter.

3.3.3 Computing method of the whole model

As the carcass and tread computing routines use the outcoming result of each other as incoming data, their coupling leads to the loop mentioned in Figure 3.3 and specified in Figure 3.16.

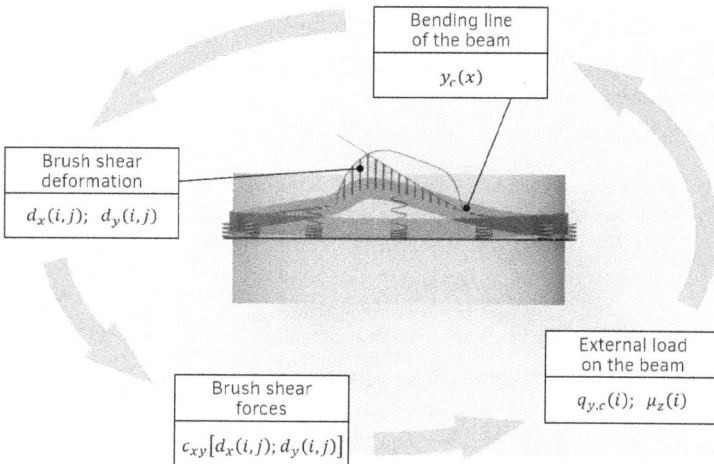

Figure 3.16. Closed loop of the complete model equations.

In this loop, the deflections and forces of two serially connected elastic elements (carcass and tread) were calculated. As tread shear stiffness is approximately one order of magnitude higher than carcass lateral stiffness, an ordinary iterative computing method was unstable and led to divergence. This issue was described in [Pac71] (Appendix A.35).

To avoid this effect, a possibility to adjust approximation step on each iteration was investigated in [Sar14] (approximation step reduction, [Kor68]). This analysis showed that this method requires low computation effort, but it does not ensure the convergence of iterative process in all motion situations. Hence, an alternative approach was developed.

Equation (3.13) for the contact part of the tire was modified: Both parts received the auxiliary summand $c_s B y_c(x)$:

$$EI y_c^{IV}(x) - T y_c''(x) + (k + c_s B) y_c(x) = q_{y.c}(x) + c_s B y_c(x) + \mu_z'(x) \quad (3.42)$$

where:

c_s – distributed shear stiffness of tread along the <u>area</u>, Equation (3.10);

B – width of the field for the contact patch, simulated in the model;

Hence: $c_s B$ – distributed shear stiffness of tread along the <u>length</u>;

$c_s B y_c(x)$ – distributed lateral force along the <u>length</u>, which together with $q_{y.c}(x)$

makes: $q_{y.c}(x_i) + c_s B y_c(x_i) = c_s B y_{ct}(i, 0.5 N_e)$. Consequently:

$$EI y_c^{IV}(x) - T y_c''(x) + (k + c_s B) y_c(x) = c_s B y_{ct}(i, 0.5 N_e) + \mu_z'(x) \qquad (3.43)$$

The motivation for this measure was a conversion of function $q_{y.c}(x)$, which greatly depends upon the carcass bending line $y_c(x)$, to function $c_s B y_{ct}(i, 0.5 N_e)$, which is independent of $y_c(x)$. This measure changed computing routines in subchapters 3.3.1-3.3.2 only in regard to insignificant details.

The bending line of a whole carcass, consisting of $y_c(x)$ in the contact part and of $y_f(x)$ in the free part of the tire, was substituted by one common function, called $y_1(x)$.

The iterative solving process is described in detail in Figure 3.17. The algorithm uses a slide-down principle:

- Firstly, the highest possible carcass deflection is calculated, in order to arrange approximation only in one direction: From the largest deflection to equilibrium deflection (Figure 3.17, steps 0.1-1.5).

- With each following iteration, the positions of contact points are "sliding down" – their lateral coordinates decrease (due to sliding) from their highest position to their equilibrium value (steps 1.8-1.12 and 2.3-2.6).

- After sliding-down, the obtained positions of the contact points were compared with their initial values and corrected, if the sliding was unnecessary for the final form of the carcass bending line (steps 2.7-2.10).

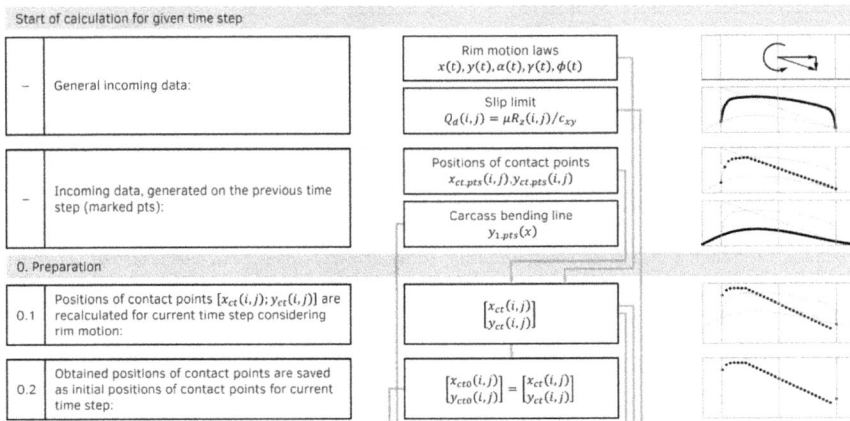

Figure 3.17. Scheme of the iterative algorithm (gray arrows – data flow, black arrows – logical flow). Part 1.

1. Initial guess

1.1	Carcass bending line $y_1(x)$ is considered to be the same as in the previous time step $y_{1.pts}(x)$:	$y_1(x)=y_{1.pts}(x)$	
1.2	Positions of root points $[x_r(i,j); y_r(i,j)]$ of the brush elements are calculated based on carcass bending line $y_1(x)$:	$\begin{bmatrix} x_r(i,j) \\ y_r(i,j) \end{bmatrix}$	
1.3	Kinematic deflection values of the tread elements d_{kx}, d_{ky} are calculated assuming absence of sliding:	$d_{kx}(i,j)$ $d_{ky}(i,j)$	
1.4	Distributed lateral force $q_{y.c}(i)$ and bending torque $\mu_z(i)$ are calculated based on kinematic deflection (assuming absence of sliding):	$q_{y.c}(i)$ $\mu_z(i)$	
1.5	For received forces a new carcass bending line $y_1(x)$ is calculated:	$y_1(x)$	
1.6	Received carcass bending line $y_1(x)$ is saved as $y_{1p}(x)$ for comparison with result of next iteration:	$y_{1p}(x)=y_1(x)$	
1.7	Positions of root points $[x_r(i,j); y_r(i,j)]$ of brush elements are updated based on new carcass bending line $y_1(x)$:	$\begin{bmatrix} x_r(i,j) \\ y_r(i,j) \end{bmatrix}$	
1.8	Kinematic deflection values of the tread elements d_{kx}, d_{ky} are calculated:	$d_{kx}(i,j)$ $d_{ky}(i,j)$	
1.9	Magnitude Q_{sk} and angle φ_s of vectors of kinematic deflections are calculated:	$Q_{sk} = \sqrt{d_{xk}^2 + d_{yk}^2}$ $\varphi_{sk} = \mathrm{atan}(d_{ky}/d_{kx})$	
1.10	Magnitude Q_s of vectors of real deflections is cut by slip limit Q_d, if necessary. Angle φ_s stays the same.	$Q_s = \min(Q_{sk}; Q_d)$ $\varphi_s = \varphi_{sk}$	
1.11	Real deflection values $d_x(i,j), d_y(i,j)$ are found by cutting the kinematic deflection values with slip limit.	$d_x(i,j) = Q_s \cos(\varphi_s)$ $d_y(i,j) = Q_s \sin(\varphi_s)$	
1.12	Positions of contact points $[x_{ct}(i,j); y_{ct}(i,j)]$ are recalculated based on real deflection values:	$\begin{bmatrix} x_{ct}(i,j) \\ y_{ct}(i,j) \end{bmatrix}$	
1.13	Distributed lateral force $q_{y.c}(i)$ and bending torque $\mu_z(i)$ are calculated based on real deflection (considering sliding):	$q_{y.c}(i)$ $\mu_z(i)$	

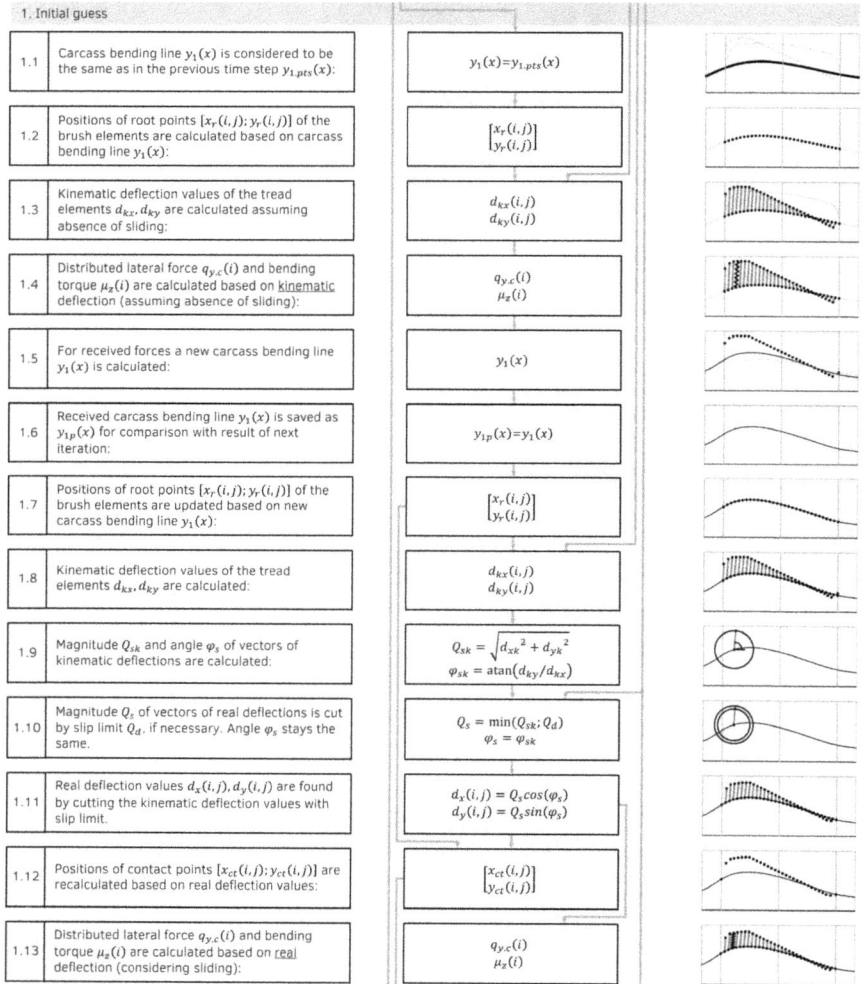

Figure 3.17. Scheme of the iterative algorithm (gray arrows – data flow, black arrows – logical flow). Part 2.

2. Second iteration

2.1	For received forces a new carcass bending line $y_1(x)$ is calculated:	$y_1(x)$	
2.2	Positions of root points $[x_r(i,j); y_r(i,j)]$ of the brush elements are updated based on new carcass bending line $y_1(x)$:	$\begin{bmatrix} x_r(i,j) \\ y_r(i,j) \end{bmatrix}$	
2.3	Kinematic deflection values of the tread elements d_{kx}, d_{ky} are calculated:	$d_{kx}(i,j)$ $d_{ky}(i,j)$	
2.4	Magnitude Q_{sk} and angle φ_s of vectors of kinematic deflections are calculated:	$Q_{sk} = \sqrt{d_{xk}^2 + d_{yk}^2}$ $\varphi_{sk} = \text{atan}(d_{ky}/d_{kx})$	
2.5	Magnitude Q_s of vectors of real deflections is cut by slip limit Q_d, if necessary. Angle φ_s stays the same.	$Q_s = \min(Q_{sk}; Q_d)$ $\varphi_s = \varphi_{sk}$	
2.6	Real deflection values $d_x(i,j), d_y(i,j)$ are found by cutting the kinematic deflection values with slip limit:	$d_x(i,j) = Q_s \cos(\varphi_s)$ $d_y(i,j) = Q_s \sin(\varphi_s)$	
2.7	Is real deflection $Q_s(i,j)$ of the i,j-th element lower than slip limit in this element?	no $\qquad$ yes $Q_s(i,j) < Q_d(i,j)$	
2.8	Magnitude Q_{s0} and angle φ_{s0} of kinematic deflection vectors between current positions of root points and initial position of contact points:	$Q_{s0} = \sqrt{(x_r - x_{ct0})^2 + (y_r - y_{ct0})^2}$ $\varphi_{s0} = \text{atan}\left(\dfrac{y_r - y_{ct0}}{x_r - x_{ct0}}\right)$	
2.9	Is real deflection $Q_s(i,j)$ of the i,j-th element lower than initial kinematic deflection $Q_{s0}(i,j)$ in this element?	no $\qquad$ yes $Q_s(i,j) < Q_{s0}(i,j)$	
2.10	Deflection of the element is restored back to the distance from its root point to initial contact point and cut by slip limit Q_d, if necessary:	$d_x(i,j) = \min(Q_{s0}, Q_d)\cos(\varphi_{s0})$ $d_y(i,j) = \min(Q_{s0}, Q_d)\sin(\varphi_{s0})$	
2.11	Positions of contact points $[x_{ct}(i,j); y_{ct}(i,j)]$ are recalculated:	$x_{ct}(i,j) = x_r(i,j) + d_x(i,j)$ $y_{ct}(i,j) = y_r(i,j) + d_y(i,j)$	
2.12	Distributed lateral force $q_{y,c}(i)$ and distributed bending torque $\mu_z(i)$ are calculated based on real deflection values (considering sliding):	$q_{y,c}(i)$ $\mu_z(i)$	
2.13	Stop criterion check: bending lines of current and previous iterations are compared: Is relative difference lower than set precision?	no $\qquad$ yes $\dfrac{y_c - y_{cp}}{y_c} < \varepsilon$	
2.14	Received carcass bending line $y_1(x)$ is saved as $y_{1p}(x)$ for comparison with result of next iteration. Iteration 2 repeats.	$y_{1p}(x) = y_1(x)$	
2.15	Forces and torque of tread are calculated based on deflection values, forces and torque of carcass are calculated based on bending line:	F_{yt}, F_{xt}, M_{zt} F_{yc}, M_{zc}	

End of calculation for given time step

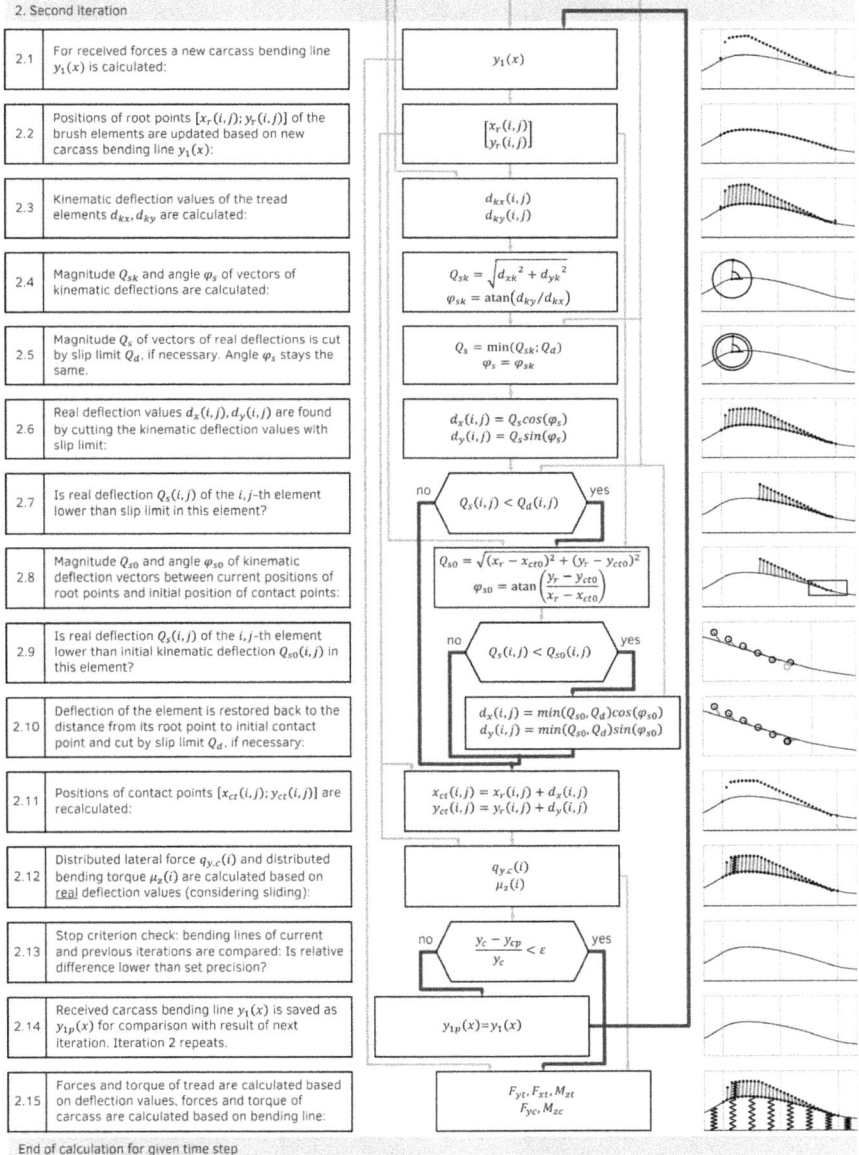

Figure 3.17. Scheme of the iterative algorithm (gray arrows – data flow, black arrows – logical flow). Part 3.

Important comments to the algorithm are following:

- The positions of the contact points $[x_{ct}(i,j); y_{ct}(i,j)]$ remain unchanged (not influenced by the static friction limit) until step 1.12.

- The slide-down algorithm changes the position of the contact points by reducing $y_{ct}(i,j)$ due to static friction limiting, but the algorithm does not increase $y_{ct}(i,j)$ if carcass deflection increases. Hence, several iterations may cause a sliding, caused by bilateral sequential approximation, which is an error. In order to prevent it, the correction is performed, which is described in steps 2.7-2.10.

Due to the slide-down approach of the algorithm, it diverges in a stable manner to a solution. However, this approach is slower than adjusted approximation step method. As long as the convergence to a solution has a higher priority, the slide-down algorithm is reasonable for the computing of model equations.

With this insight, the model description is completed. The physical mechanism has been described with a set of equations. A mathematical tool to solve them has been developed and proved. The following subchapters are devoted to preparation and analysis of this model.

3.4 Model parameterization routine

As the simulation tool had not been sufficiently developed to be suitable for commercial usage, the parameterization routine contained special, uncommon measurements for the identification of the parameters of some seldom-considered physical effects. An overview of these parameters can be found in Figure 3.11.

1. Geometrical parameters can be measured directly: free radius r and tread block height h_t.

2. With the assistance of the footprint of a stationary tire, the contact area was acquired.

    a. Contact areas, measured with different wheel loads, provided different lengths $2a_i$ and widths B_i. A polynomial interpolation was used to determine these values for continuous wheel load change;

    b. Length $2a$ and width B of the simulated contact area were selected bigger than length $2a_{max}$ and width B_{max} of the contact area, measured with the highest wheel load. Due to this measure, the model can simulate variation of the wheel load (and contact area) within a common simulation run.

    c. Pressure distribution was assumed to be uniform, reducing close to borders (Figure 3.10). Longitudinal grooves were taken into account by zeroing the pressure in the corresponding elementary areas. If measured pressure distribution is available, it can be used in the model instead of assumed uniform distribution.

3. If the friction coefficient is assumed to be temperature- and frequency-independent. It can be defined with tire lateral deflection up to full slip (Figure 2.18): The relationship between lateral-to-vertical force delivers μ. If the mentioned dependencies have to be taken into account, further measurements of friction properties are required.

4. Tensile force T (concentrated circumferential force) was calculated based on the equilibrium of the cylindrical tire shell, loaded by air pressure:

$$T = p_a B_c r \qquad (3.44)$$

where:

p_a – tire inflation pressure (gauge pressure, not absolute value);

B_c – width of the cylindrical part of the tire.

5. Carcass lateral flexibility k was defined with the help of the optical measurements of carcass deflection as a relationship between lateral force and area under the carcass bending line (Figures 2.42 and 2.27).

6. Beam bending stiffness EI and tread shear modulus G_t (or tread shear stiffness c_s) were defined with the help of measurements of tire lateral and torsional stiffness. Each of these parameters has monotone influence on tire stiffness (the higher the parameter, the higher the tire torsional and lateral stiffness). Considering the starting phase of deflection, which features practically no sliding and hence can be described with simplified models, the analytic equations for lateral and torsional stiffness can be generated, which include mentioned parameters. Due to this monotone dependence, a numerical method is appropriate to identify these parameters. The tread shear modulus, determined in such a manner (1.5 MPa), was 2.3 times higher than the measured modulus (0.65 MPa). The main reason for it is the fact that the brush model neglects the connection between neighboring elements, which can belong to a common block of the tread.

7. The special parameters are carcass shear angle coefficient k_{ps} and parameters of contact patch shape variation. The shear angle coefficient was identified for the given tire with help of the optical measurements (subchapter 2.6.3). The variation of the contact patch shape was observed with the help of acceleration measurements (subchapter 2.5).

Table 3.2. Identified values of the model parameters.

Parameter	Variable	Value	Unit
Free radius	r	0.335	m
Tread block height	h_t	0.008	m
Size of simulated field for contact area	$2a \times B$	0.125×0.220	m × m
Friction coefficient	μ	1	–
Tensile force in carcass	T	23925	N
Carcass lateral stiffness (distributed)	k	0.290	N/mm²
Carcass bending stiffness	EI	1016	Nm²
Carcass shear angle coefficient	k_{ps}	0.3	–
Tread shear modulus	G_t	1.5	MPa
Contact patch length change rate (max)	15 % per 1° of slip angle, 25 % per 1° of camber angle		

Identification of the special parameters requires very effort-consuming measurements. It would be incorrect to assume that bending behavior or contact patch shape variation do not change from tire to tire. As these effects were investigated with only one tire sample within this work, the applicability of this method for further tires cannot be properly proved without investigating the described effects with further tires. This task is the subject to future research.

3.5 Model validation

The goal of validation is to check how accurately the model represents the real system. As the aim of this model was to provide understanding of the processes, it was necessary not only to check, how accurate are the <u>output data of the model</u> (lateral force, aligning torque) in comparison to the real system, but also to check, how accurate is the <u>state of the model</u> (carcass deflection) in comparison to the real system.

The validation was performed using several different load cases (maneuvers) with the purpose to check the simulation quality of different tire properties. These maneuvers will be first introduced one by one, and then interpreted all together.

The first maneuver was a static lateral stiffness measurement: A stationary tire was displaced statically in the lateral direction (Figure 3.18). This was aimed towards the lateral properties. For two values of the measured lateral force (2 and 4 kN), plate displacement values were found (7 and 15 mm, respectively). For these displacement values, the simulated and measured carcass bending lines were compared.

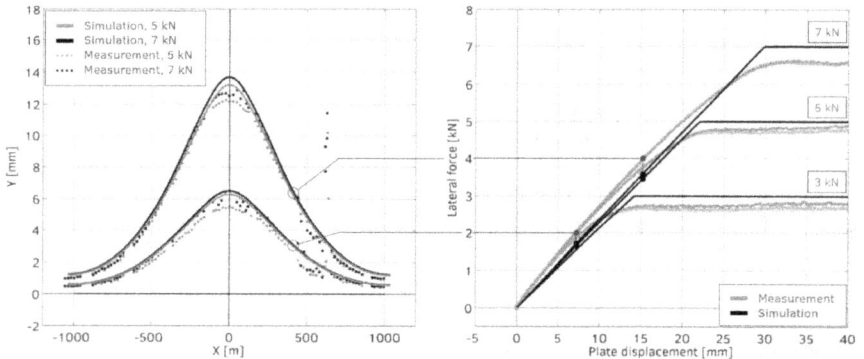

Figure 3.18. Validation with the help of the lateral stiffness measurement.

The second maneuver was the static torsional stiffness measurement: A stationary tire was turned around its vertical axis (Figure 3.19). This was aimed towards the bending properties. For three values of the measured bore torque (0.1, 0.25, and 0.4 kNm), three corresponding values of turning angle were found. Same as in the previous case, the simulated and measured carcass bending lines were compared for these displacement values.

In order to cover the non-linear transient properties, maneuvers "slip angle step" and "wheel load step" were used.

The "slip angle step" maneuver involved the following procedure: The tire was rolling at 3 km/h, the slip angle was changed stepwise from zero to a set value, kept at this value for several seconds and then changed back to zero. This maneuver made it possible to investigate three important phases:

- Excitation phase – when the tire was suddenly turned around its vertical axis;
- Saturation phase – when the tire was gaining the force and torque;
- Steady state – when the condition of equilibrium was achieved.

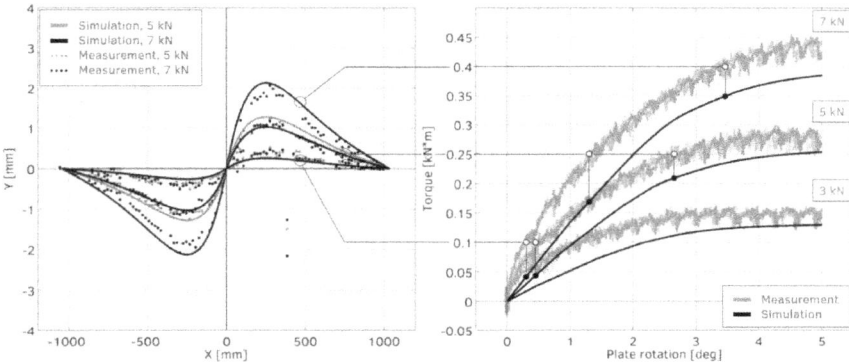

Figure 3.19. Validation using the torsional stiffness measurement.

Validation with the slip angle step maneuver was performed for different wheel load values (3-5-7 kN) and different slip angle values (1-2-3°). Qualitative correctness was also provided, but values of force and torque revealed higher errors (up to 25 %). These results can be found in Appendix A.36-A.38.

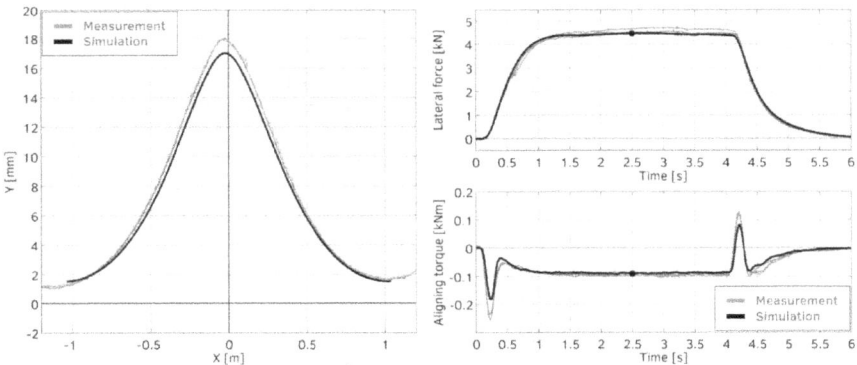

Figure 3.20. Validation using the slip angle step maneuver (wheel load of 7 kN, slip angle of 3°).

The mentioned "wheel load step" maneuver represents a tire that is rolling at 3 km/h and cornering with a slip angle of 3°. The wheel load was changed in two different sequences: 5-7-5-3-5 kN (Figure 3.21) and 7-3-7 kN (Appendix A.39).

Finally, in order to check a model response to several simultaneous or overlapping excitations, the "slip angle step" and "wheel load step" maneuvers were combined: once simultaneously (Figure 3.22) and once overlapping with a 1 s delay (Appendix A.40).

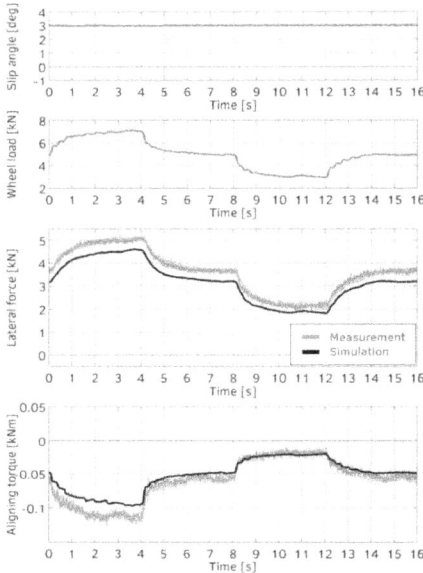

Figure 3.21. Validation using the wheel load step (5-7-5-3-5 kN, 3°).

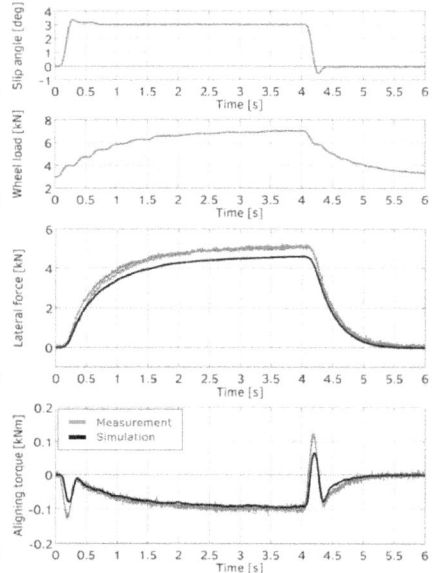

Figure 3.22. Validation using the combined simultaneous excitation (3-7-3 kN, 0-3-0°).

Validation analysis showed the following:

1. As fine tuning of the parameters was performed based on the force and torque response to the slip angle step, the highest accuracy corresponded to this maneuver (Figure 3.20).

2. Due to the limitations of the model structure, the force-displacement characteristic of the model by a lateral stiffness measurement was practically linear, whereas the real curve was not (Figure 3.18). Hence, as a result of parameterization, this curve was not tangential to the real curve at the zero point; however, it was the closest linear approximation of the real curve for the entire range of wheel load. The bending lines were here much closer to the real lines.

3. The description of lateral force in the transient process (Figure 3.20) depicts an error below 10 % for high wheel loads (5-7 kN), but in the cases of lower wheel load (3 kN) and also wheel load step, the error increases to 20 %.

4. A description of tire yaw torque can be divided into two cases:

   a. Simulation of bore torque shows generally lower values than measured results: It is visible at both torsional stiffness measurement (Figure 3.19) and slip angle step response (Figure 3.20) in excitation phases (0.0-0.4 s; 4.0-4.4 s). Simulated bending lines correlate with the measured lines with error of up to 30 %.

   b. Simulation of aligning torque was more accurate than for bore torque, also in non-steady state (Figure 3.20, 0.4-4.0 s; Figure 3.21; Figure 3.22, 0.4-4.0 s). However, error was up to 20 %.

The main reason of these deviations is limitation of brush model. Tread blocks of the real tire can be long and even continuous along the tire circumference. The tread elements that are close to the contact patch, but do not belong to it, will resist to deformation of the neighboring tread elements that are within the contact patch. Brush model is not able to consider this. Two neighboring brush elements are independent from each other. Hence, brush model shows lower lateral and torsional stiffness compared to the real tire.

5. Qualitative correctness of simulation was provided: Both the simulated output data (force and torque) and the simulated model state (carcass bending line) represented the same qualitative character as the real system.

These considerations led to the following validation conclusion:

> ! The simulation results were qualitatively correct, but quantitatively insufficiently precise. The model is therefore validated for qualitative analysis.

As the model represents a physical structure, this result is considered to be acceptable. Therefore, task 2.2 of the mission statement (p. 13) was solved. Improvement of the quantitative accuracy as well as the optimization of the computational routine will form the subject of further research.

3.6 Summary of chapter 3

Based on the set requirements and experimental observations, a suitable physical mechanism was selected. This is a combination of two simple physical subsystems: A wide flexible beam on an elastic foundation as the carcass model, and a two-dimensional array of brush elements as the tread model.

Both subsystems were physically and mathematically represented. A stable iterative algorithm for numerical solution was developed.

Parameterization of the model revealed a discrepancy of the tread shear modulus: The value that was measured with the rubber block was 2.3 times lower than the value that was identified based on measurements and the model of the tire. As long as tread material properties were not the focus of this study, background of this deviation in both modelling and measurement domains was not investigated further and is a relevant subject of future research.

Validation of the model for wide range of maneuvers showed qualitative correctness of output data (force and torque) and state data (deformation), but a lack of quantitative precision. Therefore, the developed tool can be used for qualitative analysis, providing the required understanding of the processes. This understanding is, however, limited by the assumptions and simplifications inherent in the model structure. The mentioned qualitative analysis represents the subject of the next chapter.

4 Model-based analysis

As the developed model physically represents the carcass and the tread of a tire, it provides following relevant insights:

1. Lateral and longitudinal force, aligning torque;
2. Strain state of the system, which causes the mentioned forces and torque:
    a. Strain state of the carcass: its lateral deflection and bended form;
    b. Strain state of the tread element: its deflection in longitudinal and lateral directions;
3. Physical processes and parameters that cause the mentioned strain state:
    a. Vertical load on each tread element and friction limit within it;
    b. Kinematic excitation of each tread element;
    c. Presence of sliding, sliding distance, and sliding speed (magnitude and direction).

These data clarify the effect chain of the force and torque generation, which is considered in following subchapter.

4.1 Understanding of the physical background

Strain figure in a contact patch is essential for understanding of tire response. For example, an ADAS control algorithm dynamically affects rolling wheel with slip angle (up to 35°/s [ISO11b, BreO4]) and braking torque (up to 30 Hz [Fen98]). At the same time, the wheel load, camber angle and driving torque also change dynamically, but at a slower rate (ca. 0.5-5 Hz [Hei13]). It is important to know how to manage the tire in the most efficient way in these complicated rolling conditions. Using the strain figure in the contact patch, one can not only follow the transient generation of forces and torque, but also observe what happens in the contact patch: how much friction potential is utilized in the longitudinal and lateral directions. This insight helps to understand optimal intervention on the wheel: whether to apply more brake torque or more slip angle.

The unsuitability of empirical models for this task can be illustrated with the example of the slip angle step maneuver: In the turning phase (Figure 3.19, 0.0-0.4 s), both the longitudinal and lateral forces are close to zero, but the utilized friction potential is not (due to bore torque). An identical situation occurs in the turning phase of the cornering wheel (Figure 3.19, 4.0-4.4 s). A comparison of Figure 4.1 and Figure 4.2 illustrates the deflection magnitudes in the turning phase and steady state. It explains that in the turning phase, significant friction potential can be utilized even with zero longitudinal and lateral forces (further information in Appendix A.41).

The next important issue for understanding of the physical background is the sensitivity analysis of tire behavior to its structural parameters. For example, when attempting to achieve the desired lateral vehicle performance, one must consider (among other issues) different wheel load distribution on the front and rear axles due to braking. At that point, it is helpful to know how does transient tire behavior change with different wheel load, and how this change can be influenced by tire structural parameters.

Developers of simulation tools and software for virtual testing can also profit from the mentioned understanding. As tire structure is too complicated to be reconstructed completely, a relevant question is: for a given application, which tire properties may be neglected, which ones should be considered, and how it should be done?

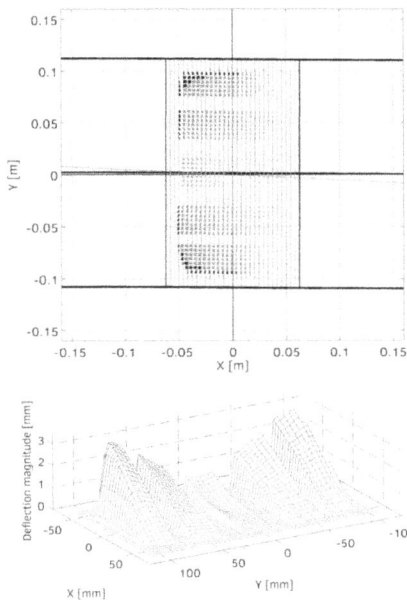

Figure 4.1. Strain figure (top) and deflection magnitude distribution (bottom) in the turning phase (Figure 3.19, 0.25 s).

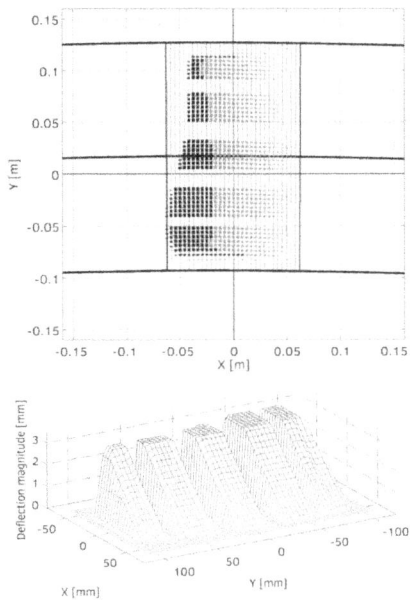

Figure 4.2. Strain figure (top) and deflection magnitude distribution (bottom) in the steady state (Figure 3.19, 2 s).

An efficient way to improve understanding of tire dynamics is an analysis of tire performance sensitivity to its structural parameters.

The parameterized model and its simulation results were used as a reference. Influencing parameters were structural parameters of the tire (e.g., tread shear stiffness) and mathematical parameters of the model (e.g., amount of brush elements). Each parameter was varied without changing the rest of the system. For three values of the parameter (reduced, reference, increased) there was performed simulation of three maneuvers: slip angle step, lateral and torsional displacement (Figures 4.3-4.4). The simulation results were evaluated using 14 criteria, which cover not only output characteristics (force and torque), but also state data (carcass deflection):

D   **Transient development of lateral force** – the rate of lateral force increase after the slip angle step.

D   **Turning peak of aligning torque** – the highest value of aligning torque after the slip angle step.

D   **Transient development of aligning torque** – the rate of aligning torque increase after the slip angle step and the peak of aligning torque.

D   **Steady-state value (lateral force, aligning torque)** – the value (of lateral force, aligning torque) in equilibrium condition after the stepwise excitation of slip angle.

D   **Carcass (highest, lowest) deflection** – the (highest, lowest) deflection of the carcass in lateral direction for a specific value of excitation.

D   **Carcass angular displacement** – orientation angle of the carcass bending line in the center of the contact patch for specific value of excitation.

D   **Linearized stiffness (lateral force, bore torque)** – the derivative (of lateral force, bore torque) in respect to (lateral, torsional) displacement at zero displacement point.

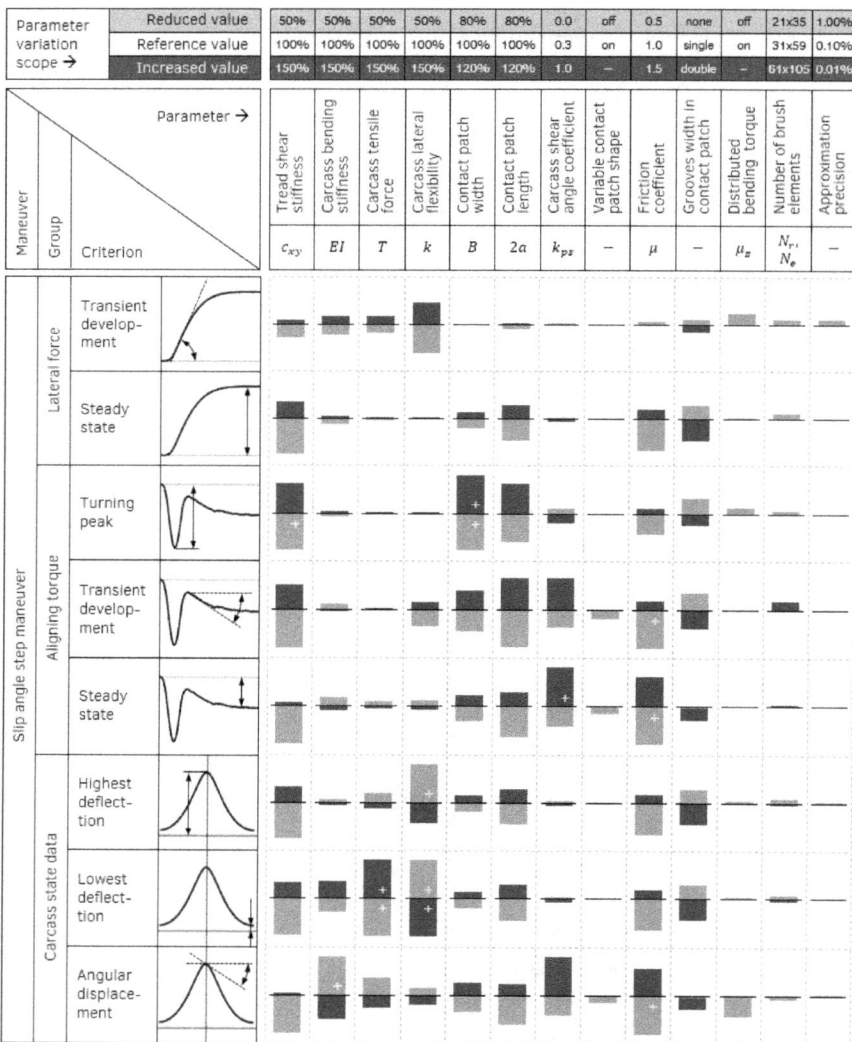

Figure 4.3. Sensitivity analysis for the given tire: Criteria of slip angle step maneuver.

Figures 4.3–4.4 illustrate these results in the following way:

- The influence of increasing a parameter on a criterion is shown with a light gray bar. The influence of reducing a parameter is depicted with a dark gray bar.
- The influence is calculated relative to the reference value of a criterion.
- If this relative value exceeded 40 %, the bar is depicted in its maximum size, and a plus sign is shown on the bar.
- Bar orientation shows the direction of influence: if dark gray bar (increased parameter) is above the axis, then the increase of the parameter led to an increase of the criterion.

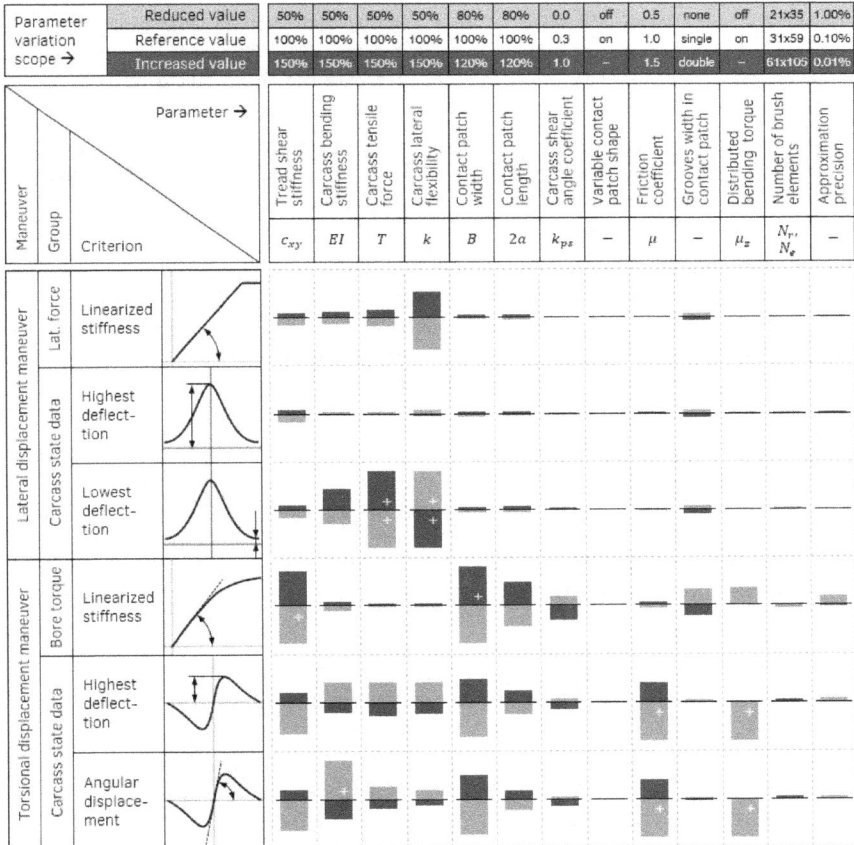

Parameter variation scope →	Reduced value	50%	50%	50%	50%	80%	80%	0.0	off	0.5	none	off	21x35	1.00%
	Reference value	100%	100%	100%	100%	100%	100%	0.3	on	1.0	single	on	31x59	0.10%
	Increased value	150%	150%	150%	150%	120%	120%	1.0	–	1.5	double	–	61x105	0.01%

Maneuver	Group	Criterion	Tread shear stiffness c_{xy}	Carcass bending stiffness EI	Carcass tensile force T	Carcass lateral flexibility k	Contact patch width B	Contact patch length $2a$	Carcass shear angle coefficient k_{ps}	Variable contact patch shape –	Friction coefficient μ	Grooves width in contact patch –	Distributed bending torque μ_z	Number of brush elements $\frac{N_r}{N_e}$	Approximation precision –
Lateral displacement maneuver	Lat. force	Linearized stiffness													
	Carcass state data	Highest deflection													
		Lowest deflection													
Torsional displacement maneuver	Bore torque	Linearized stiffness													
	Carcass state data	Highest deflection													
		Angular displacement													

Figure 4.4. Sensitivity analysis for the given tire: Criteria of lateral and torsional displacement maneuvers.

Next, there will be briefly summarized the most important insights regarding the sensitivity to the mentioned parameters. Simulation results can be found in Appendix A.42–A.54.

It is appropriate to call **tread shear stiffness** one of the most important parameters for transient handling behavior: It affects the majority of the lateral force and aligning torque criteria, and also has a large effect on carcass deflection. Stiffer tread lead to the significant increase of cornering and torsional stiffness of the tire, hence, to shorter transient phase.

The second key parameter is **carcass lateral flexibility**: Its variation significantly changes carcass deflection and is relevant for the transient development of lateral force (impact accounts for 31 %) and aligning torque (18 %).

Carcass bending stiffness influences primarily carcass angular displacement for the cornering (60 %) and torsionally-displaced tire (43 %), because higher stiffness reduces curvature of the carcass bending line. Hence, transient development of lateral force accelerates (12 %), torsional stiffness increases (7 %), but aligning torque during cornering reduces (8 %). This observation highlights that simplification of carcass bending with torsional spring cannot reasonably cover both the cornering and torsional excitation of the tire.

A very similar influence has **carcass tensile force**, as this also speeds up transient development of lateral force (10 %). An increase in tensile force reduces the bending line length, but it allows high curvature of this line. Hence, compared to carcass bending stiffness, the tensile force has a qualitatively similar, but quantitatively lower impact on carcass angular displacement (18 %).

The geometrical parameters of the contact patch also showed perceptible impact, because they change the area of the tread in the contact patch, which is proportional to the shear stiffness of the tread subsystem. An increase in **contact patch width** and an increase in its **length** both lead to the growth of torsional stiffness, steady state lateral force and aligning torque, in all three phases. However, because of the different orientation, bore torque is influenced to a greater degree by contact patch width than length, and lateral force is more affected by length than by width.

The **carcass shear angle coefficient** (Figure 4.5) has a small effect on lateral force development (4 %), a moderate impact on bore torque (16 %) and a significant influence on aligning torque in the transient phase and steady state (42 %). The physical background is rotation of the carcass cross-sections while travelling from the leading to the trailing edge (Figure 4.6). Angular displacement of the carcass is very sensitive to this coefficient (40 %). This observation proves the importance of considering the shear angle, which has been neglected in existing models [Pac12].

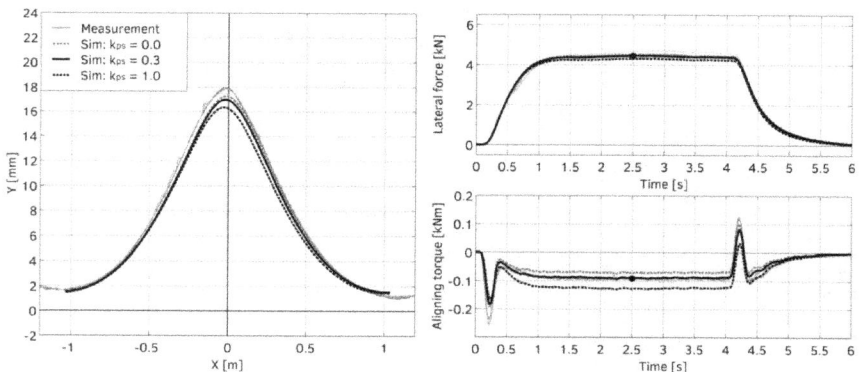

Figure 4.5. Simulation of the slip angle step maneuver with a varied carcass shear angle coefficient.

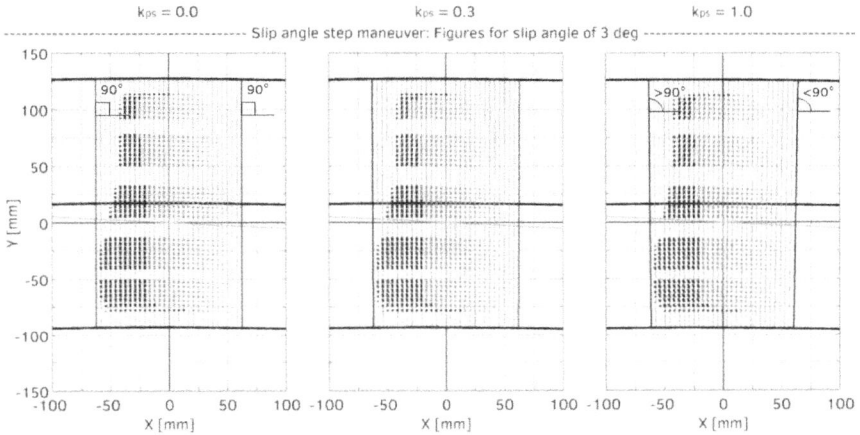

Figure 4.6. Strain figures of the slip angle step maneuver with a varied carcass shear angle coefficient.

Consideration of the variable contact patch shape has practically no effect on lateral force (1 %); however, it does have a perceptible influence on aligning torque (10 %, Figure 4.7). The highest and the lowest carcass deflection values remained almost unchanged (1 %), but the angular displacement of the carcass with a fixed contact patch shape is 7 % lower than that seen for the contact patch with a variable shape.

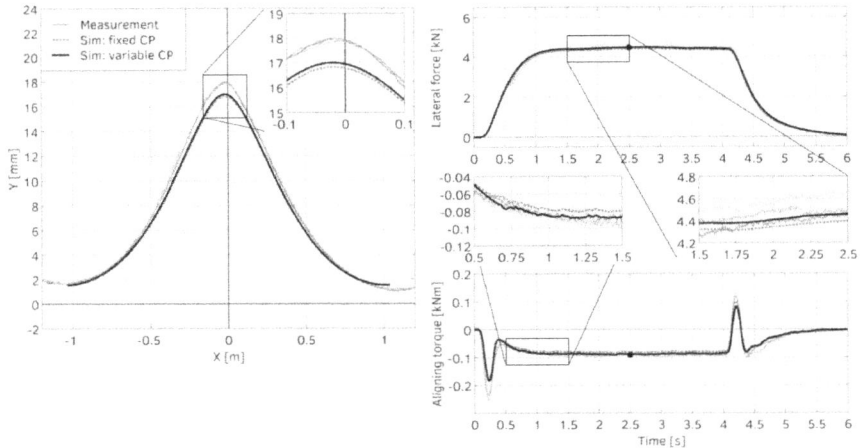

Figure 4.7. Simulation of the slip angle step maneuver with the fixed and variable contact patch shape.

The friction coefficient is only relevant in cases where slip is present. Thus, it affects the steady state force (34 %) and torque (64 %), but does not impact upon the linearized lateral and torsional stiffness or the transient development of lateral force. The influence on the transient development of the aligning torque is caused by the fact that the measurement point is close to the slip limit with a low friction coefficient. Hence, this criterion is disturbed by the slip limit.

The wider the **grooves in the contact patch**, the higher the contact pressure. Wider grooves reduce shear layer stiffness and lead to lower carcass deflection during cornering (23 %). Hence, wider grooves reduce steady state cornering force (23 %), aligning torque (15 %) and bore torque (12 %). In cases of a static lateral displacement, their influence is low (5 %), because the carcass lateral stiffness remained much lower than the shear stiffness of the grooved contact patch.

Consideration of distributed bending torque on the carcass is primarily relevant for bore torque description (18 %), because it greatly influences the calculation of carcass bending line by torsional displacement (78 %, Figure 4.8). As a consequence, the transient development of lateral force in the slip angle step maneuver is also impacted by consideration of distributed torque (11 %), as well as peak of aligning torque (6 %). This situation illustrates the importance of state data analysis (carcass deflection): Although the torque-displacement-curve in case of neglected distributed torque is closer to the measurement curve, the simulated carcass bending line in this neglecting case deviates more from the measured bending line.

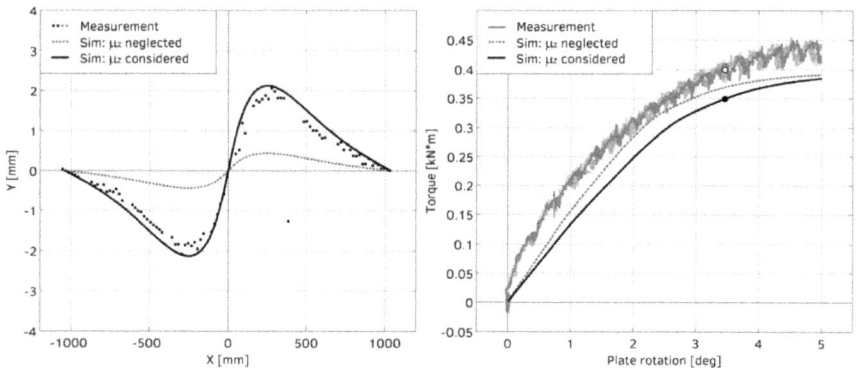

Figure 4.8. Simulation of the torsional displacement maneuver with the neglected and considered distributed bending torque.

The variation of the **number of brush elements** in the contact patch has a low influence on the results: The lower number of elements leads to criteria change below 8 % (transient development of aligning torque). Higher number of elements causes their change lower than 5 %.

A similar effect provides the variation of **approximation precision** (stop criterion of iterative solving routine): A higher error threshold (lower precision) brings criteria change below 9 % (bore torque). A lower error threshold (higher precision) causes their change of less than 2 %.

This analysis provides, on the one hand, valuable insights for tire model development (what to consider/neglect depending upon the application area). On the other hand, it clarifies the effect chain and effect interaction of the physical system of the rolling tire (which tire parameters to adjust in order to match a given criterion). These results were obtained due to the physical background of the model, which can be used for different applied tasks. One of these will be described in the following section.

4.2 An example of application

In order to illustrate the cases, in which the generated insights and tool can be useful, an example of application will be briefly introduced. A detailed description can be found in Appendix A.55.

The Finnish scientists Arto Niskanen and Ari Tuononen proposed a promising algorithm to estimate the utilization rate of the friction potential of a rolling tire [Nis17, Nis16].

> **D** **Utilized friction potential rate** – a ratio of utilized horizontal force in a contact patch to maximal static friction force.

Three acceleration sensors on a tire inner liner measured data during braking. The signal of radial acceleration clearly showed two regions: A practically constant signal in the grip region (because the sensor almost stands still) and a perceptibly oscillating signal in the slip region (because the sensor slides along a rough road surface and oscillates in the vertical direction, Figure 4.9).

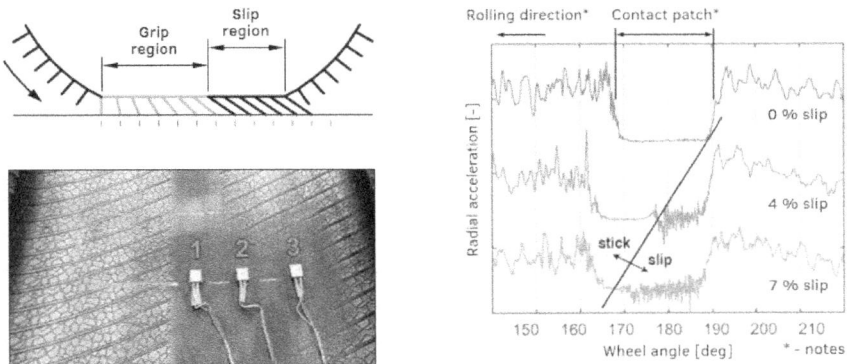

Figure 4.9. Measurement results of braking wheel by Niskanen and Tuononen [Nis17].

Hence, by means of an acceleration sensor it is possible to obtain lengths of grip and slip regions within a contact patch. This leads to two considerations, which raise two questions:

The origin of the slip is not important for the sensor.

> **?** Can this approach be used not only for braking, but also for accelerating and cornering?

Contact patch shape, and consequently the strain figure in it, may depend upon wheel load, camber angle, slip angle and angular velocity.

> **?** Can the relationship between grip and slip regions, measured in one longitudinal cross-section of the contact patch, be reliably extrapolated to the entire contact patch area?

Using the understanding gained in this study and the developed simulation model, it was possible to investigate the feasibility of the introduced method to estimate the friction potential in different rolling conditions. The comparison criterion was the utilized friction potential rate, obtained in different ways:

> **D** **Simulated utilized friction potential rate (stress based)** k_s – ratio of the sum of the <u>magnitudes</u> of the simulated shear forces of the brush elements in a contact patch to maximal friction force:

$$k_s = \frac{c_{xy} \sum_{i=1}^{N_r} \sum_{j=1}^{N_e} \sqrt{[d_x(i,j)]^2 + [d_y(i,j)]^2}}{\mu F_z} \qquad (4.1)$$

where:

$d_x(i,j)$, $d_y(i,j)$ – longitudinal and lateral deflections of the i,j-th brush element;

c_{xy} – shear stiffness of one element.

Estimated utilized friction potential rate k_e – a ratio, calculated based on the length values of grip l_g and slip l_s regions, assuming the linear development of shear stress from the leading edge to the border between the grip and slip regions.

$$k_e = 1 - \frac{l_g}{2(l_s + l_g)} \qquad (4.2)$$

Figures 4.10-4.11 depict strain in the contact patch as a top view on the tire carcass (left diagram). On their right top diagram, tire excitation is shown (slip angle, brake slip). The right bottom chart depicts the utilized friction potential rates: the simulated rate including longitudinal and lateral components, and the estimated rate. The results of friction estimation are shown for the case of using single acceleration sensor, mounted in the middle plane of the tire (equator). Hence, the estimated utilized friction potential rate was calculated based on lengths of grip and slip regions, captured on the carcass centerline.

The results highlight a limitation of this method. It is unable to capture a utilized friction potential rate below 0.5, because a perceptible slip region occurs only after the utilization of more than half of the potential. Apart from this case, estimation quality in a steady state was generally high. An error remained below 5 % independently, whether the longitudinal brake slip, longitudinal drive slip, lateral slip or combined slip (Figure 4.10) took place.

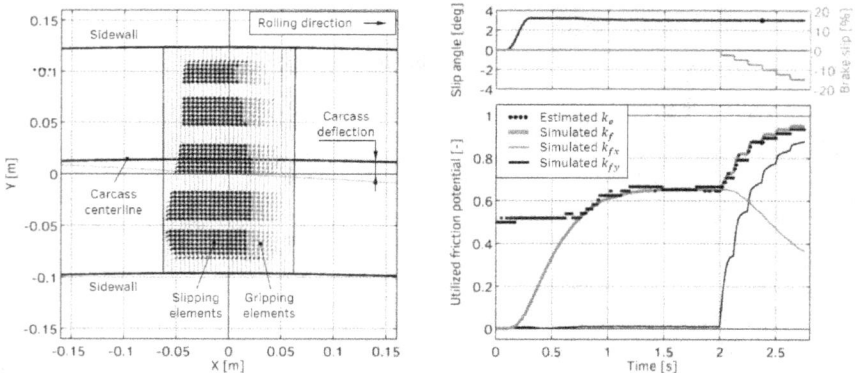

Figure 4.10. Strain figure and utilized friction potential of the cornering tire excited with brake slip (strain figure is shown for a slip angle of 3° and a brake slip of −7.5 %).

Simulation of braking excitation of a cambered tire featured trapezoidal contact patch shape and lateral slip component due to camber (Figure 4.11). Still, the estimation error of the utilized friction potential rate remained below 5 %.

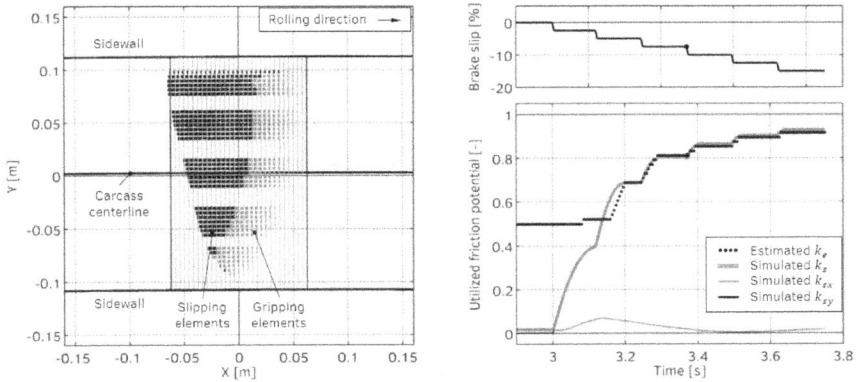

Figure 4.11. Strain figure and utilized friction potential of the cambered tire excited with brake slip (strain figure is shown for a camber angle of −4°, and a brake slip of −7.5 %).

Further details, including an analysis of the sliding speed in the brush elements in different rolling conditions, can be found in Appendix A.55. The conclusions for the current tire and measurement conditions are following:

! Although the strain figures feature asymmetry, one sensor located in the middle plane of the tire provided an estimation of the utilized friction potential rate in a steady state in the range of 0.5-1.0 with an error below 5 % for longitudinal, lateral and combined slip.

! The sliding speed in the contact patch of the cornering tire with a slip angle of 3° corresponded to the sliding speed of the straight-ahead rolling tire with a brake slip of −5 %.

Hence, application of the described estimation method is generally feasible not only for braking, but also for cornering tire. This result should not, however, be generalized for all tires without further research. It was used in this investigation only to illustrate the possible application of developed knowledge. Simulation of a tire excited with camber angle and longitudinal slip was not validated with measurements. The results of these simulation runs were also used only for the illustration of a possible application.

4.3 Summary of chapter 4

Due to the physical simulation tool with validated state description, the model-based analysis clarified the effect chains and effect interactions regarding tire transient handling behavior. It was shown, which tire structural parameters are responsible for which criteria of tire performance and how sensitive this performance is to these parameters. Influence of at least two of these parameters was not previously investigated to a sufficient extent.

A variable contact patch shape depending upon slip angle (first parameter) had practically no effect on the lateral force (1 %), but perceptibly influenced the aligning torque (10 %).

Carcass bending behavior was represented by the carcass shear angle coefficient (second parameter), which had a small influence on lateral force development (4 %), a moderate impact on the bore torque (16 %) and a significant effect on the aligning torque in the transient and steady-state phases (42 %).

Another new measure introduced in the proposed model was consideration of the distributed bending torque on the carcass. It significantly improved correctness of the carcass bending line calculation by torsional displacement (78 %). Consequently, bore torque description and transient development of lateral force during the slip angle step maneuver were also perceptibly influenced (18 % and 11 %, respectively).

The gained insights mean that the task 3.1 of the mission statement (p. 13) is accomplished.

Furthermore, it is possible to interpret these observations into general recommendations regarding consideration or neglecting of these effects in tire modelling for different applications (Table 4.1).

Table 4.1. Recommendations regarding consideration of specific physical effects for different applications.

Application / Physical effect or property	Contact patch shape change with slip angle	Shear in the carcass lateral bending	Distributed bending torque on the carcass
Simulation of vehicle lateral dynamics with steering angle or rack displacement control			
- Steady-state maneuvers	−	−	−
- Transient maneuvers	−	−	•
Simulation of vehicle lateral dynamics with steering torque control, driver-in-the-loop systems; simulation focused on steering system (steering assistance, feel, vehicle controllability, etc.)			
- Steady-state maneuvers	•	••	−
- Transient maneuvers	•	••	•
- Parking load cases	−	•	••
Legend:	− the effect may be neglected (extremely low influence)		
	• the effect should be considered (low influence)		
	•• the effect must be considered (high influence)		

It should be kept in mind though that the observations were made only with one tire and limited rolling conditions.

The functionality of the physical model made it possible to perform feasibility analysis of a new method to estimate the utilized friction potential rate of a rolling tire: This method is resistant to varying contact patch shape and can be expanded from using only longitudinal slip to also using lateral and combined slip.

In order to clarify whether and to what extent these results can be generalized for other tires, it is necessary to perform the same analysis for different types of tires. This serves as a promising subject of further research.

5 Investigation summary and discussion

5.1 Key results

The development of ADAS, autonomous driving and intelligent tire technologies brings new requirements on tire science: Transient rolling with combined slip and high slip values must be described. In order to meet increasing vehicle performance requirements and improve efficiency of vehicle development process, an understanding of the physical processes in the rolling tire must be enhanced.

A literature analysis of the existing physical models (**chapter 1**) detected issue that must be changed in order to make a model understanding-oriented instead of application-oriented. This is consideration of a wide carcass body, including its bending behavior.

Further analysis of the literature identified two relevant areas of knowledge that were not investigated deep enough yet: Contact patch shape change while cornering and carcass out-of-plane bending behavior, including the influence of these effects on tire transient handling properties.

As the method, there was selected decomposition of physical system of a rolling tire into separated physical effects and properties and estimation of their relevance for transient handling behavior. Next, there were solved three following tasks:

1. **Observation (chapter 2):**

   1.1. Contact patch shape change was captured using acceleration measurement on the tire inner liner: Shape did not perceptibly vary with rolling speed, but did change significantly with slip angle and camber angle (up to 15 % and 25 % per 1° respectively).

   1.2. Carcass bending behavior was observed using optical measurement outside and inside the tire: Bending behavior featured a significant shear angle compared to rotation angle of the cross-section (7:3). This behavior complied with the Timoshenko bending theory.

   1.3. The primary physical effects were investigated with in-tire optical measurements. Carcass lateral stiffness was displacement- and wheel-load-independent within the work range. The carcass lateral damping share of lateral force was up to 20 %.

2. **Physical analysis (chapter 3):**

   2.1. A simulation model was developed with physically justified carcass and tread description considering the observed effects. It is a combination of the wide flexible beam on an elastic foundation as the carcass model, and the two-dimensional array of brush elements as the tread model.

   2.2. Both subsystems were physically and mathematically represented. A stable iterative algorithm for numerical computing was developed.

   2.3. Validation of the model for a wide range of relevant maneuvers revealed the qualitative correctness of output data (forces) and state (strain), but a lack of quantitative precision. Hence, the developed tool can be used for qualitative analysis, providing the required understanding of the processes. However, it is limited by the assumptions and simplifications of the model structure (simplified tread material properties and friction model).

3. Understanding (chapter 4):

3.1. The model-based analysis clarified the effect chains and effect interactions regarding tire transient handling behavior. It was shown, which tire structural parameters are responsible for which criteria of tire performance and how sensitive tire performance is to different parameters.

3.2. A variable contact patch shape may be neglected in the description of transient lateral force, but it must be considered in the simulation of transient response of aligning torque, due to perceptible impact (10 %).

3.3. The presence of shear in carcass bending may be also neglected in the simulation of transient lateral force, but must also be taken into account in the aligning torque and bore torque calculation, as its impact is 16 % and 42 %, respectively.

3.4. Consideration of the distributed bending torque on the carcass, which was here introduced as a novel measure, must be considered in simulation of transient response of lateral force and bore torque (its impacts are 11 % and 18 %, respectively).

The developed tire model does not only deliver force and torque, but also shows how they are generated. The state data (carcass deflection, tread strain figure) provide an understanding of the processes. This functionality made it possible to perform feasibility analysis of a novel method to estimate the utilized friction potential rate of a rolling tire.

The model represents an open-source tool with a basic function package. On the one hand, it clarifies the limitations of a simple physical modelling approach. On the other hand, it proves that consideration of several relevant physical effects and properties can provide qualitatively correct simulation results in a wide range of tire rolling conditions. This tool can be adjusted for different applications (via neglecting irrelevant effects) as well as enhanced (via consideration of further effects or improvement of the calculation routine).

The main scientific contribution of this investigation is an improvement of the understanding of the physical background of tire transient handling behavior, which is currently needed to support vehicle development process, to increase ADAS efficiency and to improve road traffic safety.

5.2 Discussion, critique and outlook

Critical evaluation of the study reveals several limitations. These limitations can be separated into two groups: out-of-target issues and methodical weaknesses.

The model is not real-time capable (factor 1000). It does not consider tire vertical dynamics, road unevenness, deformable road structure, wet rolling. These facts are out-of-target issues, because the target of this tool does not cover these areas.

This model does not consider temperature effects, frequency-dependent friction and frequency-dependent tread material properties, although they are relevant for handling. The parameterized tread shear stiffness perceptibly deviates from the measured stiffness. These limitations were caused by the limited scope of this research: It is focused on a detailed exploration of a specific area of tire mechanics, using a simplified description of other properties. Due to the perceptible influence of frequency-dependent properties and temperature, which were simplified in this model, high-speed tests were not validated. It is therefore appropriate to consider mentioned issues in a future investigation, and analyze their influence.

The model itself contains several important methodical limitations. The carcass lateral damping was neglected. A detailed experimental investigation of this property forms the subject of further research.

Tire longitudinal stiffness is approximately two times higher than its lateral stiffness. Hence, carcass longitudinal flexibility was neglected. However, for a more precise analysis of tire rolling with longitudinal and combined slip, it is reasonable to enhance the model with this property.

The model takes into account neither mass nor gyroscopic torque. This was justified by the high stiffness of the tire. The influence of inertia compared to tire elastic forces was determined to be negligible small. However, for different tires, it would be reasonable to consider these. Due to the physical structure of the model, it can be easily accomplished.

The entire analysis was performed for one tire, one rim width and one inflation pressure value. In order to clarify whether and to what extent the gained results can be generalized for other tires, it is necessary to perform an identical analysis for different types of tires. This would serve as a promising issue for further research. The application of only one road surface is less limiting, because many of important insights of tire behavior were gained in rolling conditions with practically no slip.

The majority of measurements were performed on a drum 2 m in diameter. The physical model was both parameterized and validated with on-drum measurements. Hence, it reproduces the processes that occur in tire rolling on the drum, not on a flat surface. Thus, the research of the properties is methodically correct. However, the results should not be extrapolated to tire rolling on a flat surface without further investigation.

No validation of tread shear strain figure was performed due to the complexity of its measurement. As soon as carcass deflection is validated, the shear deformation of gripping tread blocks can be also considered to be validated (values are determined by kinematic constraints). The validation of shear deflections of sliding tread blocks is a relevant issue for future investigation.

In this research there was identified the motivation to describe and explain complex tire behavior. The first step to accomplish this is gaining of understanding of the physical effects using an understanding-oriented simulation tool. Then, there should be developed an application-oriented simulation tool for the analysis of driver assistance systems up to autonomous driving systems and tire-as-a-sensor technologies. The results of the current research did not allow to achieve the global goal. These results represent only a step on the way to this goal, but it is a necessary step to make further development possible. Such a successive exploration is the only way to succeed when working on such a demanding challenge as sustainable mobility.

References

[Ahl12]   Ahlawat, R., Smith, M. & Ichige, T., 2012. *Analyzing Tire Performance Via Wheel and Contact Patch Force Measurements.* Cologne, Tire Technology Expo 2012.

[Amm97]   Ammon, D., Gipser, M., Rauh, J. & Wimmer, J., 1997. High Performance System Dynamics Simulation of the Entire System Tire-Suspension-Steering-Vehicle. *Vehicle System Dynamics,* Volume 27, pp. 435-455.

[AUD15]   AUDI AG. (2015). Presskit 08/15. Retrieved from: https://www.audi-mediacenter.com/de

[Bäc12]   Bäcker, M. & Gallrein, A., 2012. *CDTire: State-of-the-art Tire Models for Full Vehicle Simulation.* Detroit, Americas HyperWorks Technology Conference.

[Bak91]   Bakker, E. & Pacejka, H. B., 1991. The magic formula tyre model. *Proceedings of 1st Colloquium Tyre Models for Vehicle Dynamics Analysis,* pp. 1-18.

[Bel00]   Belkin, A. et al., 2000. Dynamischer Kontakt des Radialreifens als viskoelastische Schale mit einer starren Stützfläche bei stationärem Rollen. *Technische Mechanik,* 20(4), pp. 355-372.

[Bel97]   Belkin, A., Bukhin, B., Mukhin, O. & Narskaya, N., 1997. Some models and methods of pneumatic tire mechanics. *Vehicle System Dynamics: International Journal of Vehicle Mechanics and Mobility,* Volume 27, pp. 250-271.

[Böh66]   Böhm, F., 1966. Mechanik des Gürtelreifens. *Ingenieur Archiv,* Band 35, p. 82.

[Böh66b]   Böhm, F., 1966. *Zur Mechanik des Luftreifens,* Technische Hochschule Stuttgart: Habilitationsschrift.

[Böh85]   Böhm, F., 1985. *Theorie schnell veränderlicher Rollzustände für Gürtelreifen.* S.l., Springer-Verlag, pp. 30-44.

[Böh88]   Böhm, F., Eichler, M. & Kmoch, K., 1988. *Grundlagen der Rolldynamik von Luftreifen.* Essen, Fortschritteder Kraftfahrzeugtechnik 1. Fachtagung Fahrzeug-Dynamik. Haus der Technik, pp. 3-34.

[Bre04]   Breuer, B. & Bill, K. H., 2004. *Bremsenhandbuch: Grundlagen, Komponenten, Systeme, Fahrdynamik.* 2 ed. Wiesbaden: Springer-Verlag.

[Bre95]   Breuer, W., 1995. Ragmomentenregelung bei PKW, Düsseldorf: Forschr.-Ber. VDI-Reihe 12, Nr. 235.

[Bro25]   Broulhiet, G., 1925. *La suspension de la direction de la voiture automobile. Schimmi et dandinement.* Bul. 78. P. 12.: Société des ingéniers civils de France.

[Cal14]   Calabrese, F., Bäcker, M. & Gallrein, A., 2014. *Advanced Handling Applications with New Tire Model Utilizing 3D Thermo-Dynamics.* S.l., SIMPACK User Meeting.

[Cal15]   Calabrese, F., Bäcker, M. & Gallrein, A., 2015. Evaluation of different modeling approaches for the tire handling simulations – analysis and results. *Proceedings of 6th International Munich Chassis Symposium 2015 Chassis.Tech plus,* pp. 749-773.

[Cal15b]   Calabrese, F., Bäcker, M. & Gallrein, A., 2015. *A Method to Combine an MBD Tire Model with a Thermo-dynamical one to improve the accuracy in the tire simulations.* Barcelona, ECCOMAS Thematic Conference on Multibody Dynamics.

[DEK15]   DEKRA Automobil GmbH, 2015. *Road Safety Report 2015. A future based on experience*, Stuttgart: DEKRA Automobil GmbH.

[Don89]   Donges, E., Auffhammer, R. & Fehrer, P., 1989. Aktive Hinterachskinematik (AHK) – Neue Entwicklungsmöglichkeiten in der Fahrzeugquerdynamik. *VDI Berichte,* Issue 778, pp. 265-283.

[Ein10]   Einsle, S., 2010. *Analyse und Modellierung des Reifenübertragungsverhaltens bei transienten und extremen Fahrmanövern,* s.l.: Dissertation, TU Dresden.

[Ell69]   Ellis, J. R., 1969. *Vehicle Dynamics.* Advanced School of Automotive Engineering, Cranfield: London Business Books Limited.

[Erd09]   Erdogan, G., Alexander, L. & Rajamani, R., 2009. *Wireless piezoelectric sensor for the measurement of tire deformations and the estimation of slip angle.* Hollywood, California, Proceedings of the ASME 2009 Dynamic Systems and Control Conference.

[Erd11]   Erdogan, G., 2011. Novel wireless tire deformation sensors for estimation of tire slip angle and tire-road friction coefficient. *Twin Cities: University of Minnesota, Department of Mechanical Engineering.*

[ETRO6]   The European Tyre and Rim Technical Organisation, 2006. *ETRTO Standards Manual.* Brussels: s.n.

[Fal69]   Falk, S., 1969. *Lehrbuch der Technischen Mechanik. Dritter Band: Die Mechanik des elastischen Korpers.* Berlin/Heidelberg: Springer Verlag.

[Fen98]   Fennel, H., 1998. ABS plus und ESP – Ein Konzept zur Beherrschung der Fahrdynamik. *ATZ – Automobiltechnische Zeitschrift,* 100(4), pp. 302-308.

[Feo99]   Feodesjev, V., 1999. *Strength of materials.* Moscow: Bauman Moscow State Technical University.

[Fév07]   Février, P. & Fandard, G., 2007. A new thermal and mechanical tire model for handling simulation. *VDI-Berichte,* Volume 2014.

[Fév08]   Février, P. & Fandard, G., 2008. Thermal and mechanical tire modelling to simulate handling. *Automobiltechnische Zeitschrift (ATZ) ,* Issue 05.

[Fév10]   Fevrier, P., Martin, H. & Fandard, G., 2010. *Method for simulating the thermomechanical behavior of a tire rolling on the ground.* New York, Patent No. US 2010/0010795 A1.

[Fia54]   Fiala, E., 1954. Seitenkräfte am rollenden Luftreifen. *VDI Zeitschrift,* Band 96, pp. 973-979.

[Fro41]   Fromm, H., 1941. *Kurzer Bericht über die geschichte der Theorie des Radflatterns.* S.l., NACA TM 1365, pp. 19-41.

[Gar16]   Garcia, M. & Kaliske, M., 2016. *A consistent implementation for inelastic materials in an ALE formulation for steady state rolling contact.* Braunschweig, GAMM – Joint Annual Meeting.

[Gie12]   Gießler, M., 2012. *Mechanismen der Kraftübertragung des Reifens auf Schnee und Eis.* Karlsruhe: Karlsruher Schriftenreihe Fahrzeugsystemtechnik.

[Gim01]   Gim, G. & Choi, Y., 2001. Role of tire modeling on the design process of a tire and vehicle system. 2001 Korea ADAMS User Conference, 11, 8-9.

[Gim07]   Gim, G., Choi, Y. & Kim, S., 2007. A semi-physical tire model for a vehicle dynamics analysis of handling and braking. *Vehicle System Dynamics,* Issue 45, p. 169-190.

[Gip97]   Gipser, M., Hofer, R. & Lugner, P., 1997. Dynamical Tyre Forces Response to Road Unevennesses. *Vehicle System Dynamics ,* Volume 27, pp. 94-108.

[Goo17]   Goodyear Dunlop Tires, 2017. *Our 124-year history.* [Online]
             Available at: http://www.dunloptires.com/en-US/company/tire-history#1888-1922
             [Accessed 26.01.2017].

[Han13]   Hanada, R. et al., 2013. A Sampling Moire Method to Measure the Dynamic Shape and Strain of Rotating Tires. *Tire Science and Technology, TSTCA,* 41(3), pp. 214-225.

[Har11]   Hartmann, B., 2011. *Emergency Brake and Steer Assist. Ein integriertes Fahrerassistenzsystem für Notsituationen.* Vienna, 9. ÖAMTC Symposium Reifen und Fahrwerk.

[Har13]   Harrer, M., Görich, H.-J., Reuter, U. & Wahl, G., 2013. *50 years 911 – the perfecting of the chassis.* Munich, Chassis.tech plus – 4th International Munich Chassis Symposium.

[Hei13]   Heißing, B., Ersoy, M. & Gies, S., 2013. *Fahrwerkhandbuch: Grundlagen, Fahrdynamik, Komponenten, Systeme, Mechatronik, Perspektiven.* 4 ed. Wiesbaden: Springer Vieweg.

[Hig97]   Higuchi, A., 1997. *Transient Response of Tyres at Large Wheel Slip and Camber,* Delft: Delft University of Technology.

[Höf01]   Höfer, P., Kaliske, M. & Thiele, K., 2001. *Vorhersage von Reifenkennlinien mit FEM Simulation.* Hannover, VDI-Tagung Reifen-Fahrwerk-Fahrbahn.

[Hoo05]   de Hoogh, J., 2005. *Implementing inflation pressure and velocity effects into the magic formula tyre model,* Eindhoven: Eindhoven University of Technology.

[Hol99]   Holdmann, P., Köhn, P. & Ammon, D., 1999. Das Einlaufverhalten von Reifen in unterschiedlichen Betriebssituationen und seine Relevanz für die Gesamtfahrzeugdynamik. *VDI Berichte,* Issue 1494, pp. 139-155.

[ISO11]   ISO 3888-2:2011, 2011. *Passenger cars – Test track for a severe lane-change manoeuvre – Part 2: Obstacle avoidance.* Genève: International Organization for Standardization.

[ISO11b]   ISO 7401:2011, 2011. *Road vehicles – Lateral transient response test methods – Open-loop test methods.* Genève: International Organization for Standardization.

[ISO99]   ISO 3888-1:1999, 1999. *Passenger cars – Test track for a severe lane-change manoeuvre – Part 1: Double lane-change*. Genève: International Organization for Standardization.

[Jag15]   Jager, B. et al., 2015. *Torque-Vectoring Stability Control of a Four Wheel Drive Electric Vehicle*. Seoul, 2015 IEEE Intelligent Vehicles Symposium (IV).

[Jal10]   Jalali, K., 2010. *Stability Control of Electric Vehicles with In-Wheel Motors*, Waterloo: University of Waterloo.

[Jo13]   Jo, H. Y. et al., 2013. Development of Intelligent Tire System. *SAE Technical Paper*, 2013-01-0744.

[Joh17]   Johnson, B., kein Datum *Robert William Thomson. Historic UK – History Magazine*. [Online] Available at:http://www.historic-uk.com/HistoryUK/HistoryofScotland/Robert-William-Thomson/ [Accessed 26.01.2017].

[Kal10]   Kaliske, M., 2010. *Numerical Modeling in Tire Mechanics*. Bamberg, 9. LS-DYNA Forum.

[Kel12]   Kelly, D. & Sharp, R., 2012. Time-optimal control of the race car: influence of a thermodynamic tyre model. *Vehicle System Dynamics: International Journal of Vehicle Mechanics and Mobility*, 50(4), pp. 641-662.

[Kim15]   Kim, S., Kim, K.-S. & Yoon, Y.-S., 2016. Development of a tire model based on an analysis of tire strain obtained by an intelligent tire system. *International Journal of Automotive Technology*, 16(5), pp. 865-875.

[Kno06]   Knoll, P. M. & Langwieder, K., 2006. *Der Sicherheitseffekt von ESP in Realunfällen – Überlegungen zum volkswirtschaftlichen Nutzen von prädiktiven Fahrerassistenzsystemen*. Garching, Tagungsband der Tagung „Sicherheit durch Fahrerassistenz". TUM/TÜV Süd.

[Kor68]   Korn, G. A. & Korn, T. M., 1968. *Mathematical Handbook for Scientists and Engineers*. New York: McGraw Hill Book Company.

[Kuw13]   Kuwayama, I., Matsumoto, H. & Heguri, H., 2013. *Development of a next-generation-size tire for eco-friendly vehicles*. München, 4. Internationales Münchner Fahrwerk-Symposium Chassis.Tech plus.

[Kuw14]   Kuwayama, I., Matsumoto, H. & Heguri, H., 2014. *Development of Large and Narrow Tire Technology as "Ologic"*. Aachen, 23rd Aachen Colloquium Automobile and Engine Technology 2014.

[Köh06]   Köhn, P., Richter, T., Smakman, H. & Vieler, H., 2006. Integrated Chassis Management – ein Weg zur Integrierten Fahrdynamikregelung. *15. Aachener Kolloquium Fahrzeug- und Motorentechnik*, pp. 775-792.

[Kva06]   Kvasnicka, P. et al., 2006. Durchgängige Simulationsumgebung zur Entwicklung und Absicherung von Fahrdynamischen Regelsystemen. *VDI-Berichte*, Issue 1967, pp. 387-403.

[Lan15]   Langen, P., 2015. *The all new BMW 7 series*. Munich, Chassis.tech plus – 6th International Munich Chassis Symposium.

[Mas15]   Masago, T., 2015. *A Development of Tire Wear Estimation with the use of Acceleration of a Tire.* Dresden, 11th Intelligent Tire Technology Conference.

[Mat15]   Matilainen, M. & Tuononen, A., 2015. Tyre contact length on dry and wet road surfaces measured by three-axial accelerometer. *Mechanical Systems and Signal Processing,* Issue 52-53, pp. 548-558.

[Nis14]   Niskanen, A. & Tuononen, A. J., 2014. Three 3-Axial Accelerometers Fixed Inside the Tyre for Studying Contact Patch Deformations in Wet Conditions. *Vehicle System Dynamics,* 52(5), pp. 287-298.

[Nis15]   Niskanen, A. & Tuononen, A. J., 2015. Three Three-Axis IEPE Accelerometers on the Inner Liner of a Tire for Finding the Tire-Road Friction Potential Indicators. *Sensors,* 15(8), pp. 19251-63.

[Nis16]   Niskanen, A. J., Xiong, Y. & Tuononen, A. J., *Towards the Friction Potential Estimation: A Model-Based Approach to Utilizing In-Tyre Accelerometer Measurements.* 2016 IEEE Intelligent Vehicles Symposium, Gothenburg.

[Nis17]   Niskanen, A. & Tuononen, A., 2017. Detection of the local sliding in the tyre-road contact by measuring vibrations on the inner liner of the tyre. *Measurement Science and Technology,* 28(5), pp. 1-12.

[Pac12]   Pacejka, H. B., 2012. *Tire and Vehicle Dynamics.* 3 ed. Oxford: Elsevier Ltd.

[Pac71]   Pacejka, H. B., 1971. *A hybrid computer model of tire shear force generation,* s.l.: Document 3. Automobile Manufacturers Association, Highway Safety Research Institute.

[Pop93]   Popp, K. & Schiehlen, W., 1993. *Fahrzeugdynamik.* Stuttgart: Teubner Verlag.

[Sak95]   Sakai, E., 1995. Measurement and Visualization of the Contact Pressure Distribution of Rubber Disks and Tires. *Tire Science and Technology, TSTCA,* 4(23), pp. 238-255.

[Sar14]   Sarkisov, P., Prokop, G. & Popov, S., 2014. *Reifenmodellierung für den instationären Betrieb durch Kombination der Funktionsansätze.* RWTH Aachen, 1. WKM Symposium.

[Sar15]   Sarkisov, P., Kubenz, J., van Putten, S. & Prokop, G., 2015. Steifer Reifenprüfstand für höhere Messgenauigkeit. *ATZ Automobiltechnische Zeitschrift,* 4, pp. 46-51.

[Sar16]   Sarkisov, P., Prokop, G. & Wensch, J., 2016, 16/1. Solving the linear inhomogeneous differential equation in the simulation model of passenger car tire. *PAMM – Proceedings in Applied Mathematics and Mechanics. Joint 87th Annual Meeting of the International Association of Applied Mathematics and Mechanics (GAMM) and Deutsche Mathimatiker Vereinigung (DMV),* pp. 233-234.

[Saw96]   Sawase, K. et al., 1996. Development of the active yaw control system. JSAE, 50(11), pp. 52-57.

[Sch05]   Schmeitz, A., Besselink, I., de Hoogh, J. & Nijmeijer, H., 2005. *Extending the Magic Formula and SWIFT tyre models for inflation pressure changes,* s.l.: TU Eindhoven.

[Sel14]   Selig, M. et al., 2014. Rubber Friction and Tire Dynamics: A Comparison of Theory with Experimental Data. *Tire Science and Technology, TSTCA,* 42(4), pp. 216-262.

[Soc05]   Société de Technologie Michelin, 2005. *Der Reifen. Haftung – was Auto und Straße verbindet.* Clermont-Ferrand: Michelin Reifenwerke KgaA.

[Sta14]   Stappen, H.-J., 2014. *Reply from Dr. Ing. h.c. F. Porsche AG to Teknikens Värld magazine.* [Online] Available at: http://teknikensvarld.se/porsche-macan-behaves-strangely-in-the-moose-test-162276/ [Accessed 06.02.2015].

[Tib13]   Tiberio, E. & Morinaga, H., 2013. *Bridgestone's CAIS Technologie für eine detaillierte Fahrbahnzustandsklassifizierung.* 11. ÖAMTC Symposium, "Reifen und Fahrwerk", Vienna.

[TNO13]   TNO Automotive, 2013. *MF-Tyre/MF-Swift 6.2. Help Manual,* s.l.: Delft-tyre.

[Tor15]   Torbrügge, S., 2015. *Testing and Understanding of Tire-Road Interaction on Wet Roads.* 13. ÖAMTC Symposium "Reifen und Fahrwerk", Vienna.

[Uil07]   Uil, R., 2007. *Tyre models for steady-state vehicle handling analysis,* Eindhoven: Eindhoven University of Technology.

[Unr13]   Unrau, H.-J., 2013. *Der Einfluss der Fahrbahnoberflächenkrümmung auf den Rollwiderstand, die Cornering Stiffness und die Aligning Stiffness von Pkw-Reifen.* Karlsruhe: Karlsruher Schriftenreihe Fahrzeugsystemtechnik.

[vPu12]   van Putten, S., 2012. *Tire Bore Torque. A semi-mechanical approach to measurement and modeling of standstill and rolling conditions.* Intelligent Tire Technology Conference, Darmstadt.

[vPu17]   van Putten, S., 2017. *Eine hybride Methode zur objektiven Beschreibung von Reifencharakteristika,* Dissertation, Technische Universität Dresden.

[Wag17]   Wagner, A. & van Putten, S., 2017. *Audi chassis development – Attribute based component design.* Wiesbaden, Springer Vieweg.

[WHO15]   World Health Organization, 2015. *Global status report on road safety 2015,* Geneva: WHO Press.

[Wie08]   Wiesenthal, M., Collenberg, H. F. & Krimmel, H., 2008. *Aktive Hinterachskinematik AKC - ein Beitrag zu Fahrdynamik, Sicherheit und Komfort.* 17. Aachener Kolloquium Fahrzeug- und Motorentechnik 2008, Aachen.

[Wit11]   Wittenburg, J. & Pestel, E., 2011. *Festigkeitslehre: Ein Lehr- und Arbeitsbuch.* 3 ed. Berlin: Springer-Verlag.

[Wun03]   Wunderlich, W. & Pilkey, W. D., 2003. *Mechanics of Structures. Variational and Computational Methods.* 2 ed. s.l.: CRC Press LLC.

[Yim15]   Yim, S., 2015. Unified chassis control with electronic stability control and active front steering for under-steer prevention. *International Journal of Automotive Technology,* 16(5), pp. 775-782.

[Yu01]   Yu, Z.-X., Tan, H.-F., Du, X.-W. & Sun, L., 2001. A Simple Analysis Method for Contact Deformation of Rolling Tire. *Vehicle System Dynamics,* 36:6, pp. 435-443.

List of abbreviations

ABS	Anti-lock Brake System	[Fen98]
ADAS	Advanced Driver Assistance System	[Kno06]
ARP	Active Rollover Prevention	[Sta14]
ESA	Emergency Steering Assist	[Har11]
ESP	Electronic Stability Program (also ESC)	[Fen98]
ETRTO	European Tyre and Rim Technical Organisation	[ETR06]
FEM	Finite Element Method	[Kal10]
FFT	Fast Fourier Transform	[Sar16]

List of symbols

a	Half of the length of the largest contact patch
$a_{i0,1}$	Coefficients of linear approximation
a_n	Coefficients of a Fourier series
A_{shear}	Shear area of a tread element
b	Half of the length of the free circular part of a tire
b_n	Coefficients of a Fourier series
B_c	Width of the cylindrical part of the tire
B	Width of the largest contact patch
c_s	Shear stiffness of tread, distributed along the area
c_{shear}	Shear stiffness of a tread element
c_{xy}	Shear stiffness of the tread fragment, which corresponds to one brush element
C	Geometrical contact patch center
$CX_C Y_C Z_C$	Contact coordinate system
$C_{c.i.j}$	Constant coefficients of integration in a fragment of the contact part of tire
$C_{c1...4}$	Constant coefficients of integration in the whole contact part of tire
$C_{f1...4}$	Constant coefficients of integration in the free part of tire
ds	Elementary length of the beam
dt	Simulation time step

dx	A projection of the elementary length of the beam onto longitudinal axis
$d_k(i,j)$	Magnitude of kinematic deflection of the i,j-th brush element
$d_{kx}(i,j)$	Longitudinal kinematic deflection of the i,j-th brush element
$d_{ky}(i,j)$	Lateral kinematic deflection of the i,j-th brush element
$d_x(i,j)$	Longitudinal real deflection of the i,j-th brush element
$d_y(i,j)$	Lateral real deflection of the i,j-th brush element
$\vec{e_1}$	Basis vector, oriented along the beam axis
EI	Bending stiffness of a beam in the given cross-section
G_t	Shear modulus
h_t	Height of a tread element
i	Index of a brush element in longitudinal direction
j	Index of a brush element in lateral direction
k	Stiffness coefficient of elastic foundation
k_e	Estimated utilized friction potential rate
k_f	Simulated utilized friction potential rate, force based
$k_{fx}\ (k_{fy})$	Simulated utilized friction potential rate in longitudinal (lateral) direction, force based
k_m	Measured utilized friction potential rate
$k_{mx}\ (k_{my})$	Measured utilized friction potential rate in longitudinal (lateral) direction
k_{ps}	Carcass shear angle coefficient
k_s	Simulated utilized friction potential rate (stress based)
$k_{sx}\ (k_{sy})$	Simulated utilized friction potential rate in longitudinal (lateral) direction, stress based
l_z	Width of the area, which corresponds to one brush element
l_g	Length of the grip region in a contact patch
l_s	Length of the slip region in a contact patch
l_x	Length of the area, which corresponds to one brush element
$\vec{M}$	Vector of internal concentrated torque in a beam cross-section
$M_{fc1,2}$	Matrices of coefficients
M_z	Internal concentrated bending torque in a beam cross-section
N_e	Number of brush elements in one row in the contact patch (width)
n_h	Number of harmonic components of a Fourier series
N_r	Number of brush elements in one line in the contact patch (length)

$OX_WY_WZ_W$	Wheel coordinate system
$O_RX_RY_RZ_R$	Road coordinate system
p_a	Tire inflation pressure (gauge)
$\vec{q}$	Vector of external distributed force on a beam
q_i	Discretized right-hand side function of bending equation
q_x	Longitudinal external distributed force on a beam
q_y	Lateral external distributed force on a beam
$q_{y.c}(x)$	Distributed load of tread shear force on a beam
$q_{y.f}(x)$	Distributed load of flexible force of the elastic foundation on a beam
$q_{y.t}(x)$	Distributed load of longitudinal tensile force on a beam
$\vec{Q}$	Vector of internal concentrated force in a beam cross-section
Q_y	Lateral internal concentrated force in a beam cross-section
r	Free radius of an inflated tire
$R_{fc1,2}$	Vectors of fixed terms
$R_z(i,j)$	Vertical load on the i,j-th brush element
T	Concentrated longitudinal tensile force in a cross-section of a beam
x	Longitudinal coordinate of a beam
$x_o(t)$	Longitudinal coordinate of the rim center
$x_{ct}, x_{ct0}(i,j)$	Longitudinal position of the contact point of the i,j-th brush element
x_i	Node point coordinate between intervals on a beam
$x_r(i,j)$	Longitudinal position of a root point of a brush element
$y(x)$	Function of the lateral deflection of a beam
$y_1(x)$	Function of the lateral deflection of the entire carcass
$y_o(t)$	Lateral coordinate of the rim center
$y_c(x)$	Function of the lateral deflection of the carcass in the contact part of a tire
$y_{c.i}(x)$	Solution of the bending equation for the i-th interval
$y_{c.ih.i}(x)$	Inhomogeneous part of the solution of the bending equation
$y_{ct}, y_{ct0}(i,j)$	Lateral position of the contact point of the i,j-th brush element
$y_f(x)$	Function of the lateral deflection of the carcass in the free part of a tire
$y_r(i,j)$	Lateral position of a root point of a brush element
$z_o(t)$	Vertical coordinate of the rim center
$\alpha(t)$	Wheel slip angle

$\gamma(t)$	Wheel camber angle
Δx_{shear}	Shear deflection of a tread element
θ	Cross-section orientation angle
$\theta(t)$	Wheel roll angle
$\lambda_{c1,3}$	Solutions of the characteristic polynomial of the bending equation in the contact part of a tire
$\lambda_{f1,3}$	Solutions of the characteristic polynomial of the bending equation in the free part of a tire
$\vec{\mu}$	Vector of external distributed torque on a beam
μ_z	Magnitude of the external distributed bending torque on a beam
μ	Friction coefficient
$\varphi_k(i,j)$	Orientation angle of the vector of kinematic deflection of the i,j-th brush element
$\varphi(t)$	Wheel yaw angle
$\phi(t)$	Wheel rotation angle
$\psi(t)$	Wheel pitch angle
$\omega(t)$	Wheel rotation speed

List of tables

List of figures

Appendix

Table of content

A.1. Experimental equipment: Hydraulic pulsing rig

This test bench is able to apply a load in one direction with a high frequency (Table A.1). The test sample can be preloaded in the second direction with the help of a pneumatic cylinder. The rig is equipped with a thermographic camera to monitor warmth development. Due to high-frequent excitation, precise sensors and high stiffness, the pulsing machine met the demands of this investigation.

Table A.1. Hydraulic pulsing machine technical data.

Max. excitation force	±50 kN
- Linearity	0.08 % of full scale
- Hysteresis	0.05 % of full scale
- Repeatability	0.03 % of full scale
Max. preload force	6.2 kN
- Accuracy	0.5 % of full scale
Displacement range	±125 mm
- Accuracy	0.01 mm
Excitation frequency	up to 100 Hz

A.2. In-tire acceleration measurement: Reliability analysis

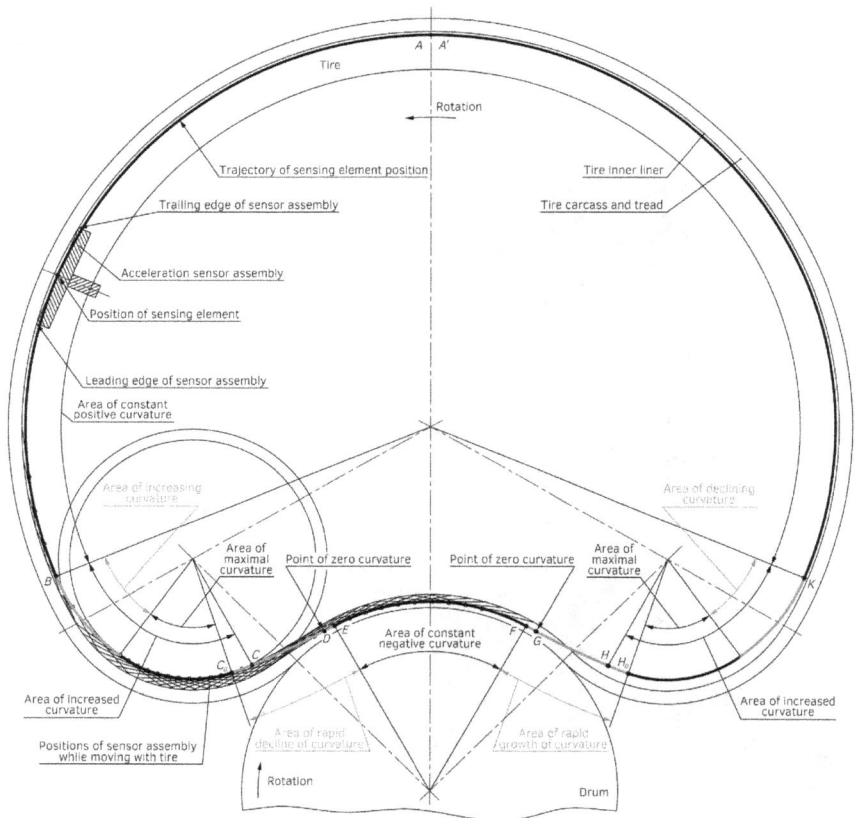

Figure A.2.1. Simplified tire-drum interaction (dimensions are exaggerated for better visibility).

The measured radial acceleration (Figure A.2.1) provided two reliable conclusions:

1. Zone AC_0 and zone H_0A' corresponds to the free part of the tire;
2. Zone EF corresponds to the contact patch.

Consequently, the start of the contact patch is between points C_0 and E; the end of the contact patch is between points F and H_0. Quantitatively, for the current example, the length was between 56 mm and 116 mm. The reason for such a high deviation is the extended zones of acceleration drop (C_0E = 31 mm) and rise (FH_0 = 29 mm). In order to obtain information about the contact patch length, it was first necessary to understand the background of the radial acceleration drop and rise.

In [Yu01] a simulation model was developed, which described tire radial deformation around the contact patch. The carcass was represented as an inextensible ring on elastic foundation with damping. Further, in [Nis16], radial acceleration was measured and simulated using a similar model. Both the simulation model and the measurements with an acceleration sensor showed that the value of acceleration decreased and increased stepwise (instead of ramps observed in this research).

! For typical conditions, the tire outer surface changes its curvature in an almost stepwise manner, and this occurred directly at the border points of the contact patch.

! The acceleration sensor assembly, which was used in this research, smoothed the rapid acceleration drop and rise due to its high length and high stiffness.

In order to provide stability and quality of the data evaluation, the points C, D, G, and H were used for evaluation, because their positions could be detected with the smallest influence from noise. For each experiment, the results of all revolutions, performed with the same parameters, were analyzed together. Then, three repetitions of the experiment were also considered together: The positions of one point (e.g., C), detected in different revolutions and different experiment repetitions, are depicted as a histogram (Figure A.2.2). Additionally, this figure shows a footprint of a non-rolling tire on the same drum with the same wheel load.

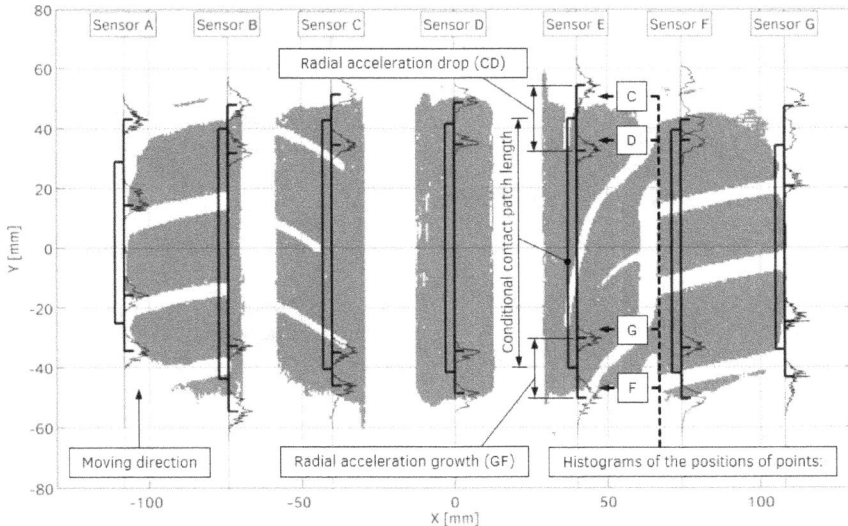

Figure A.2.2. Comparison of the footprint measurement of a stationary tire and the measurement of the contact patch length with acceleration sensors (example).

The following subchapters depict measurement results with seven scatter diagrams and one footprint figure each. Scatters illustrate the positions of the conditional start and end of the contact patch. The light grey, dark grey and black color correspond to three repetitions of the same measurement.

Leaps between the values are caused by the lateral grooves between the tread blocks. The sensors A, B, G, and F were mounted on the tread tracks, which were interrupted by lateral grooves. As the contact length on these tracks increases, the acceleration sensor was able to measure it only stepwise, with a noticeable leap in the scatter.

A.3. Rolling speed variation: Tire, 3 kN, camber 0 deg

Test type: Rolling speed variation
Test object: Tire
Wheel load: 3 kN
Rolling speed: 20 - 120 km/h
Slip angle: 0 deg
Camber angle: 0 deg

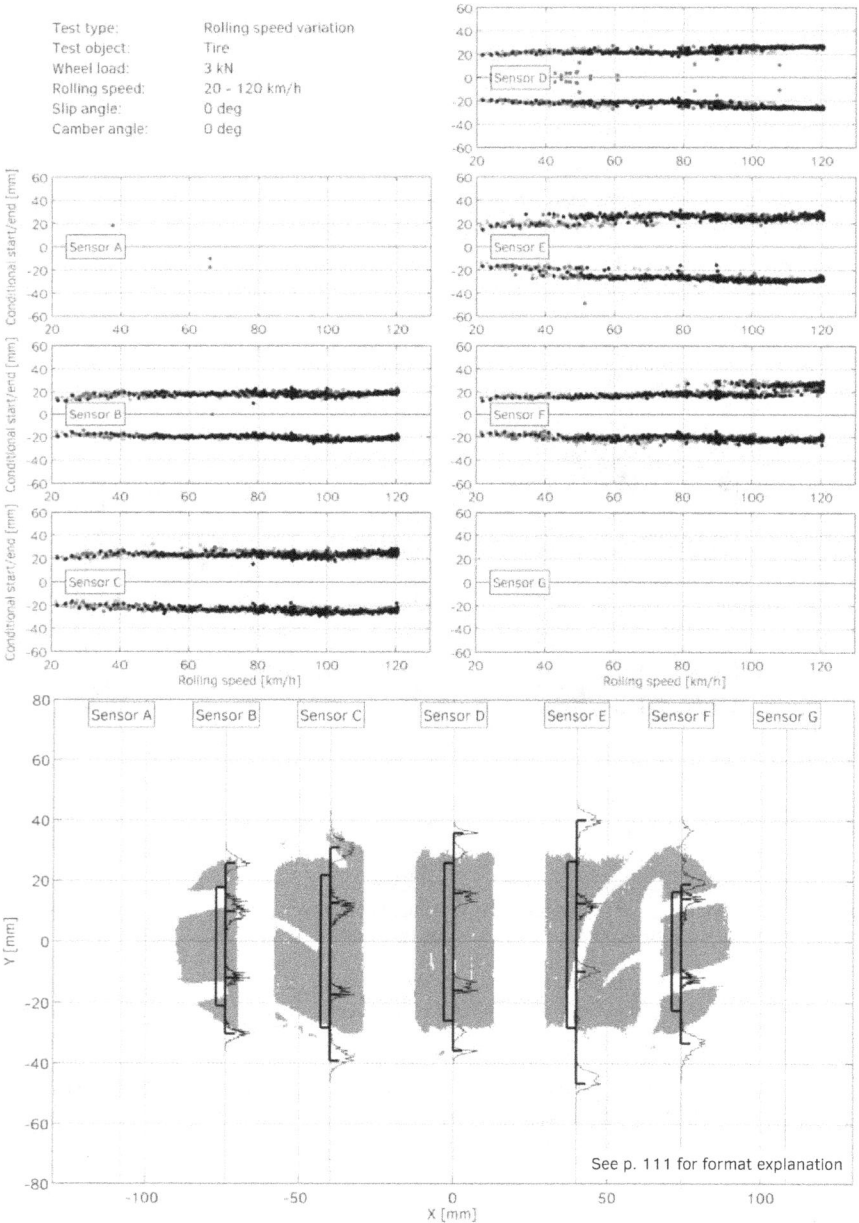

See p. 111 for format explanation

A.4. Rolling speed variation: Tire, 3 kN, camber –4 deg

Test type:	Rolling speed variation
Test object:	Tire
Wheel load:	3 kN
Rolling speed:	20 - 120 km/h
Slip angle:	0 deg
Camber angle:	–4 deg

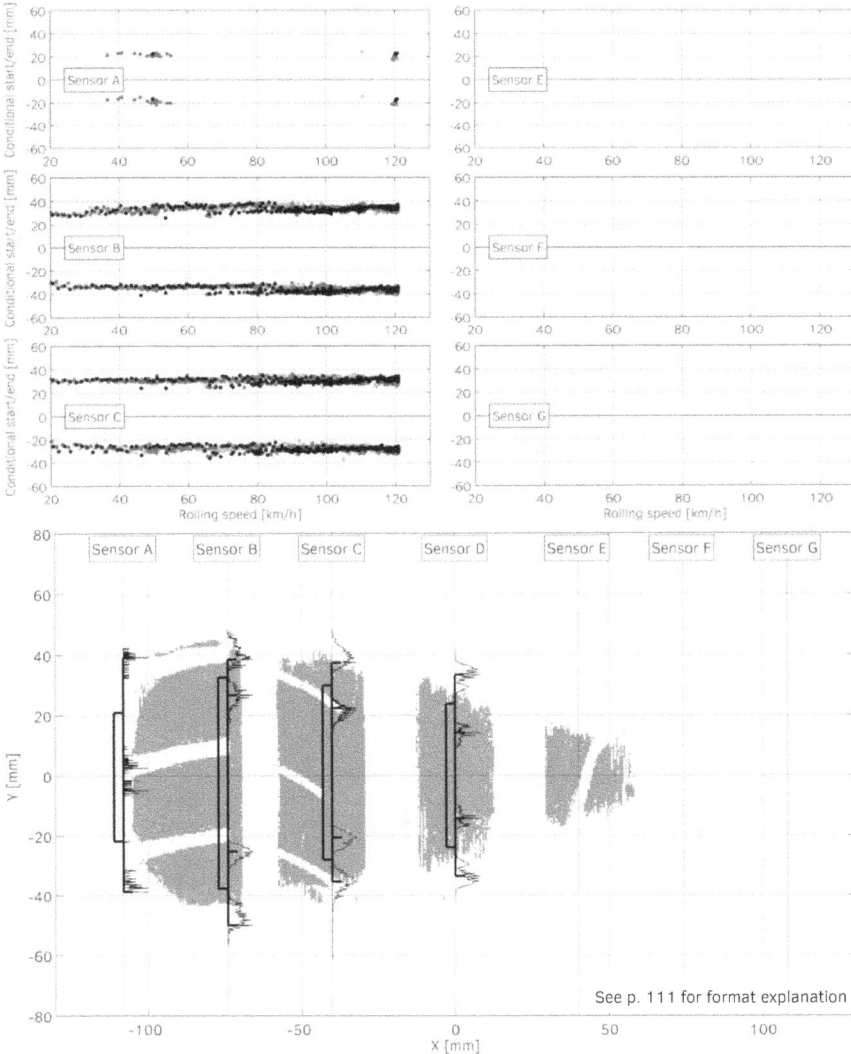

See p. 111 for format explanation

A.5. Rolling speed variation: Tire, 5 kN, camber 0 deg

Test type: Rolling speed variation
Test object: Tire
Wheel load: 5 kN
Rolling speed: 20 - 120 km/h
Slip angle: 0 deg
Camber angle: 0 deg

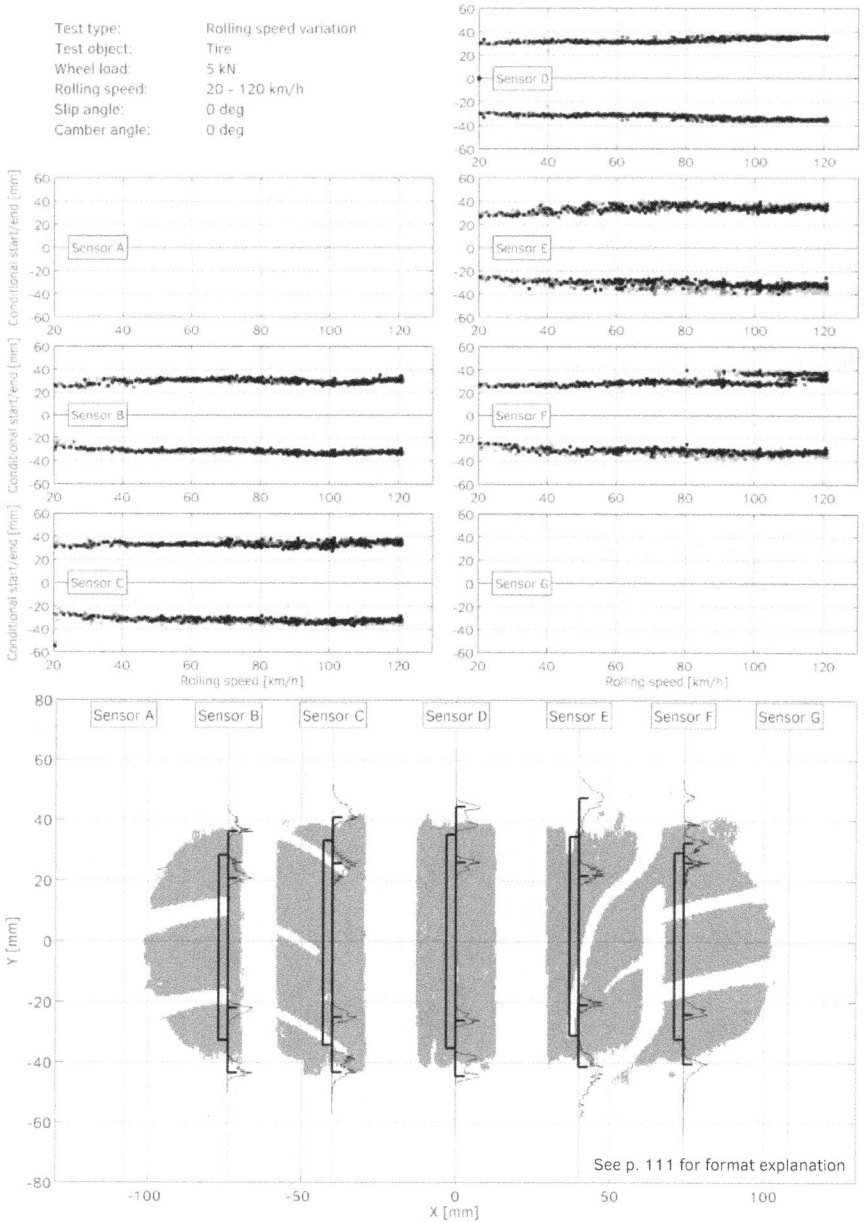

See p. 111 for format explanation

A.6.   Rolling speed variation: Tire, 5 kN, camber −4 deg

Test type:	Rolling speed variation
Test object:	Tire
Wheel load:	5 kN
Rolling speed:	20 – 120 km/h
Slip angle:	0 deg
Camber angle:	−4 deg

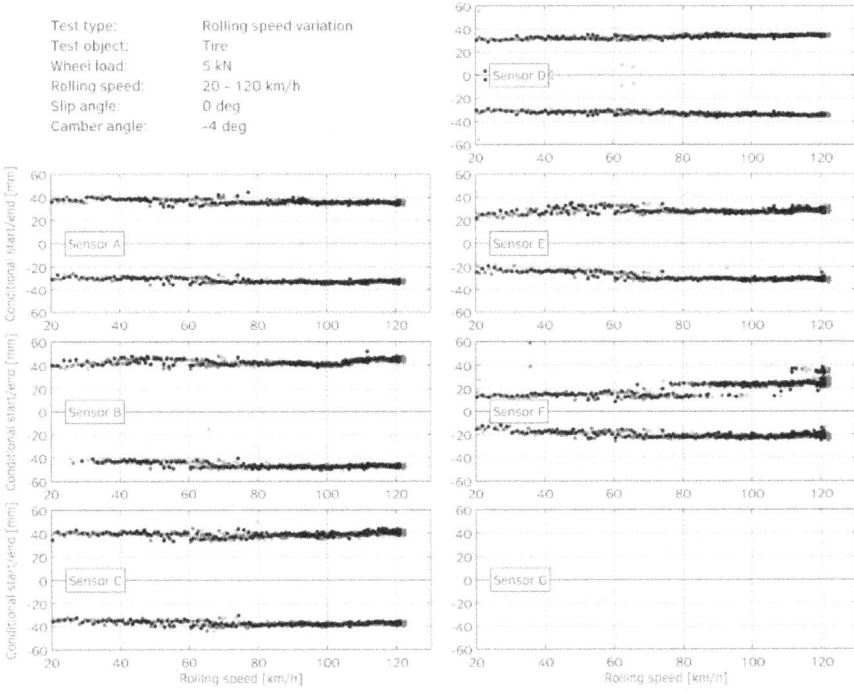

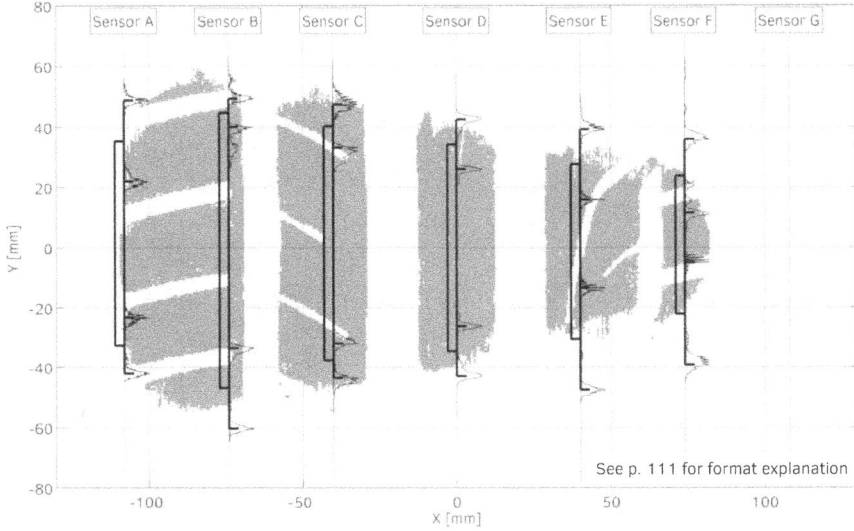

See p. 111 for format explanation

A.7.   Rolling speed variation: Tire, 7 kN, camber 0 deg

Test type:	Rolling speed variation
Test object:	Tire
Wheel load:	7 kN
Rolling speed:	20 - 120 km/h
Slip angle:	0 deg
Camber angle:	0 deg

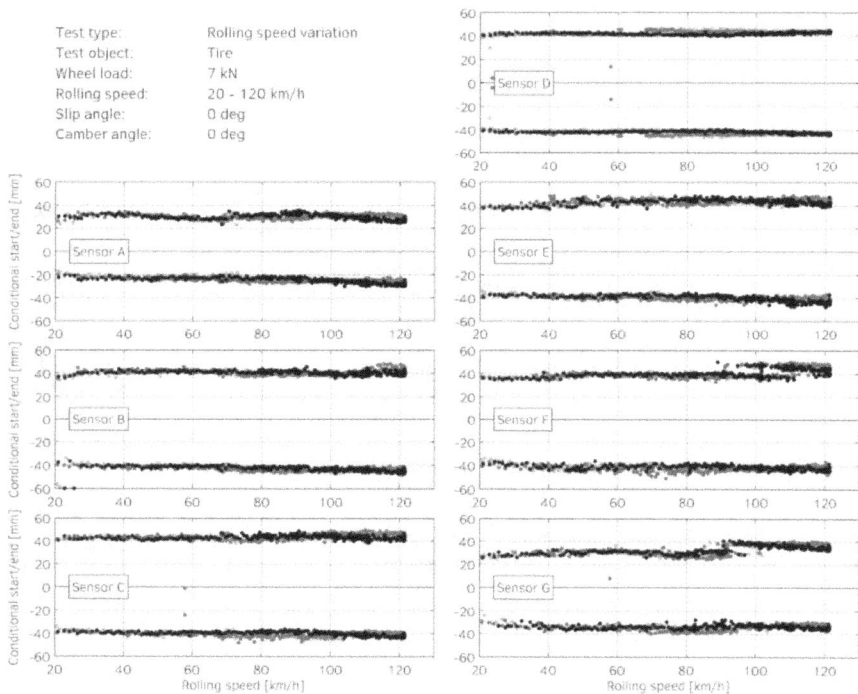

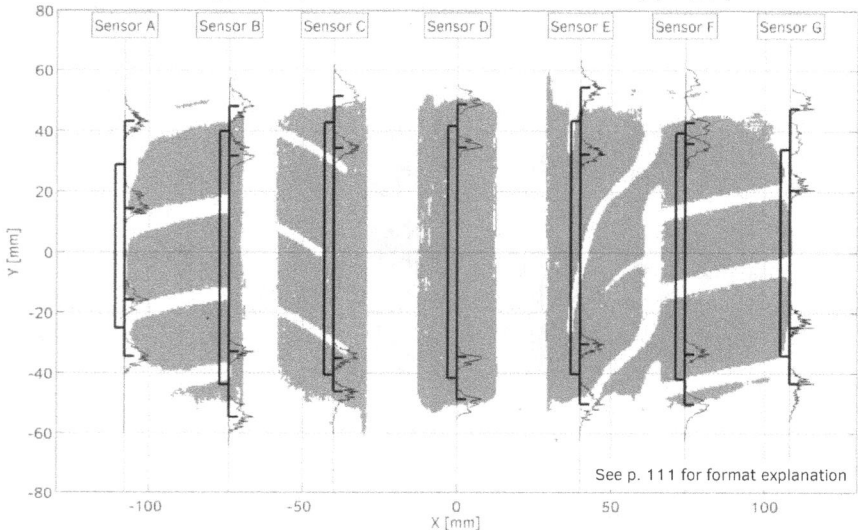

See p. 111 for format explanation

A.8.   Rolling speed variation: Tire, 7 kN, camber –4 deg

Test type: Rolling speed variation
Test object: Tire
Wheel load: 7 kN
Rolling speed: 20 - 120 km/h
Slip angle: 0 deg
Camber angle: -4 deg

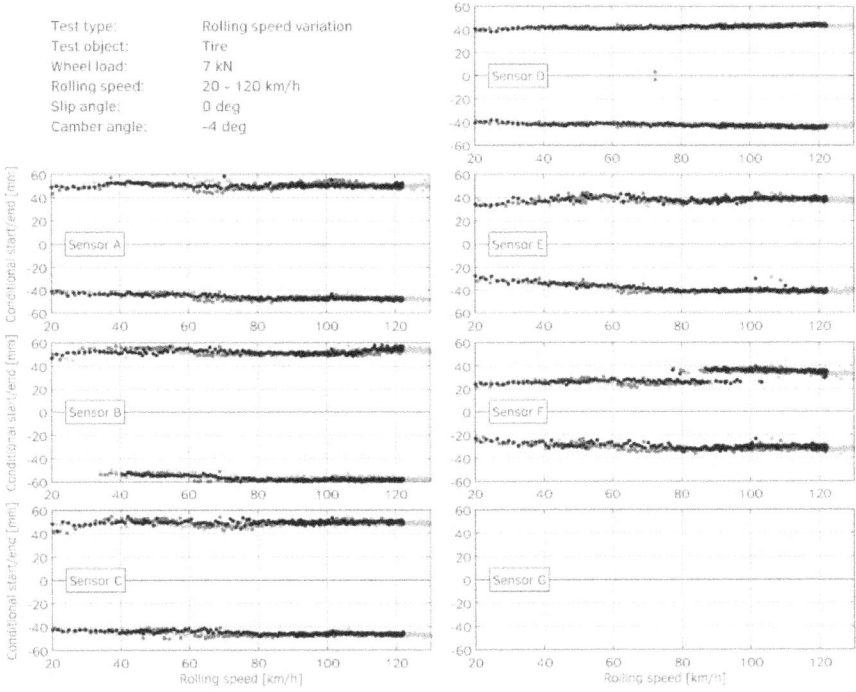

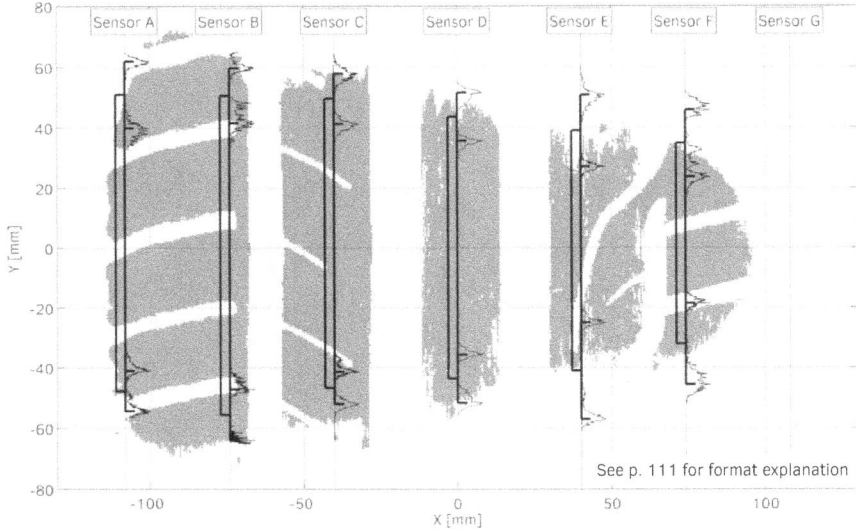

See p. 111 for format explanation

A.9. Rolling speed variation: Carcass, 3 kN, camber 0 deg

Test type: Rolling speed variation
Test object: Carcass
Wheel load: 3 kN
Rolling speed: 20 - 100 km/h
Slip angle: 0 deg
Camber angle: 0 deg

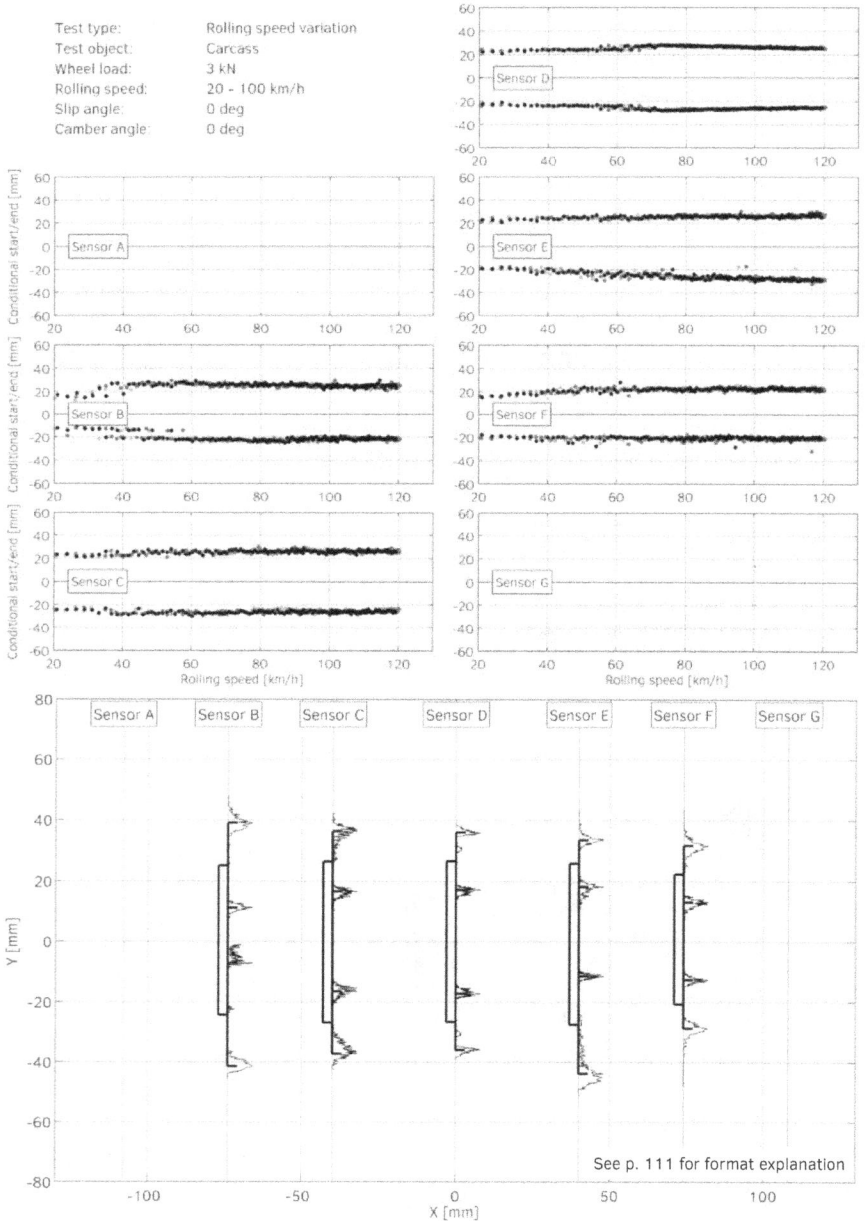

See p. 111 for format explanation

A.10. Rolling speed variation: Carcass, 3 kN, camber −4 deg

Test type:	Rolling speed variation
Test object:	Carcass
Wheel load:	3 kN
Rolling speed:	20 - 100 km/h
Slip angle:	0 deg
Camber angle:	−4 deg

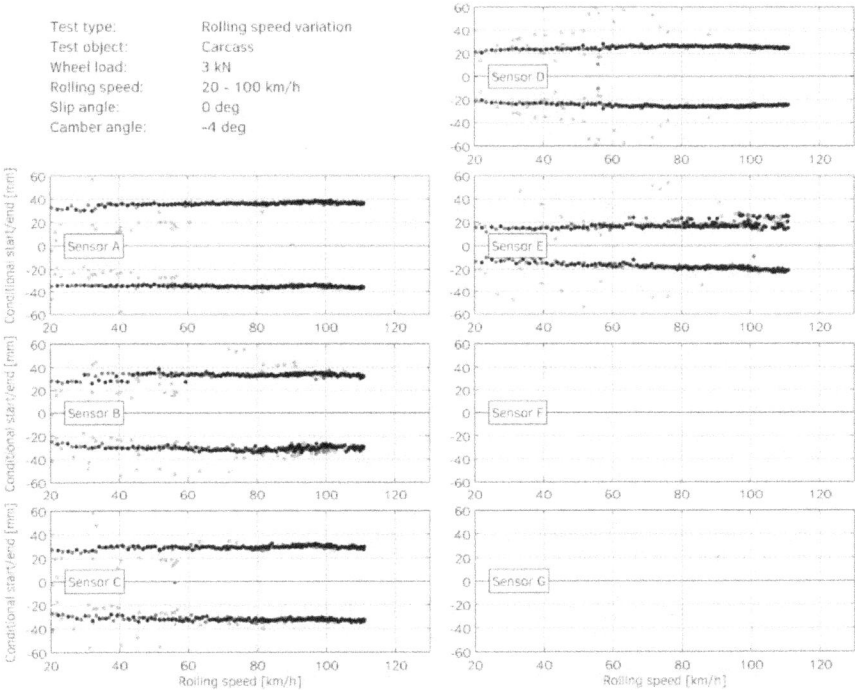

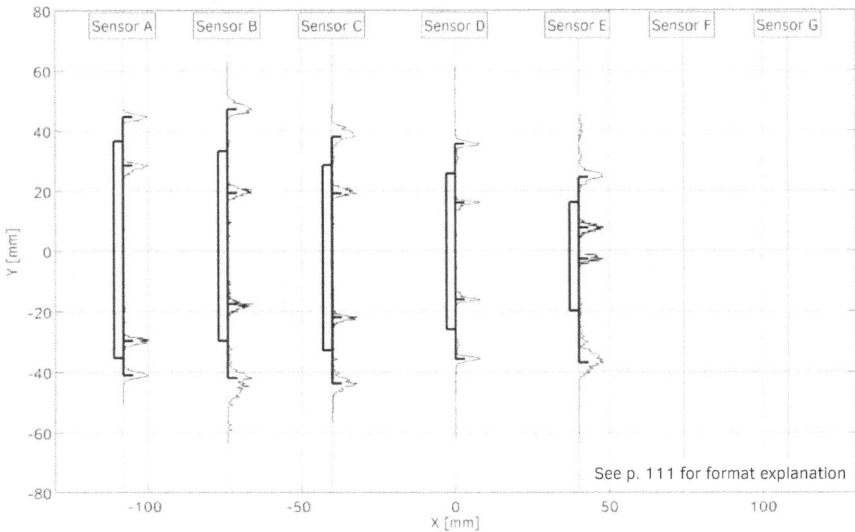

See p. 111 for format explanation

A.11. Rolling speed variation: Carcass, 5 kN, camber 0 deg

Test type: Rolling speed variation
Test object: Carcass
Wheel load: 5 kN
Rolling speed: 20 - 100 km/h
Slip angle: 0 deg
Camber angle: 0 deg

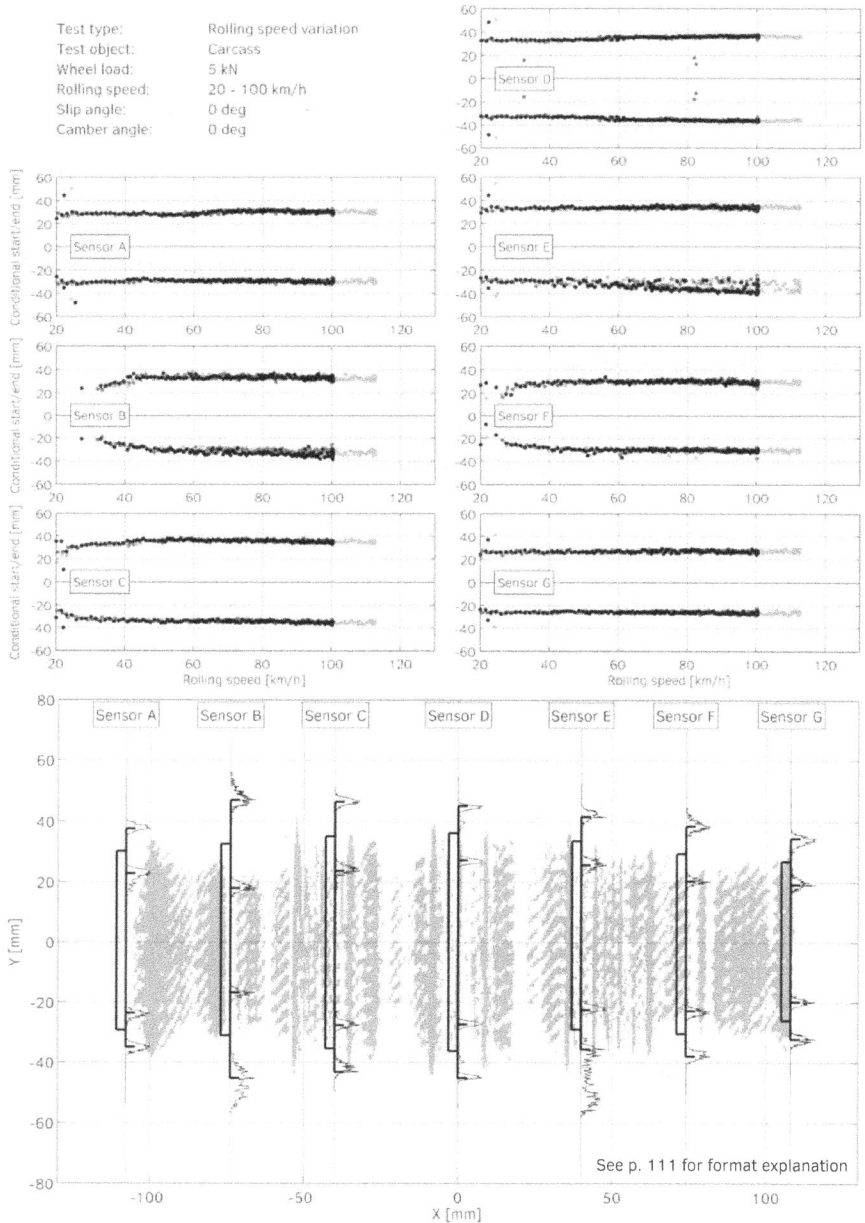

See p. 111 for format explanation

A.12. Rolling speed variation: Carcass, 5 kN, camber −4 deg

Test type:	Rolling speed variation
Test object:	Carcass
Wheel load:	5 kN
Rolling speed:	20 – 100 km/h
Slip angle:	0 deg
Camber angle:	−4 deg

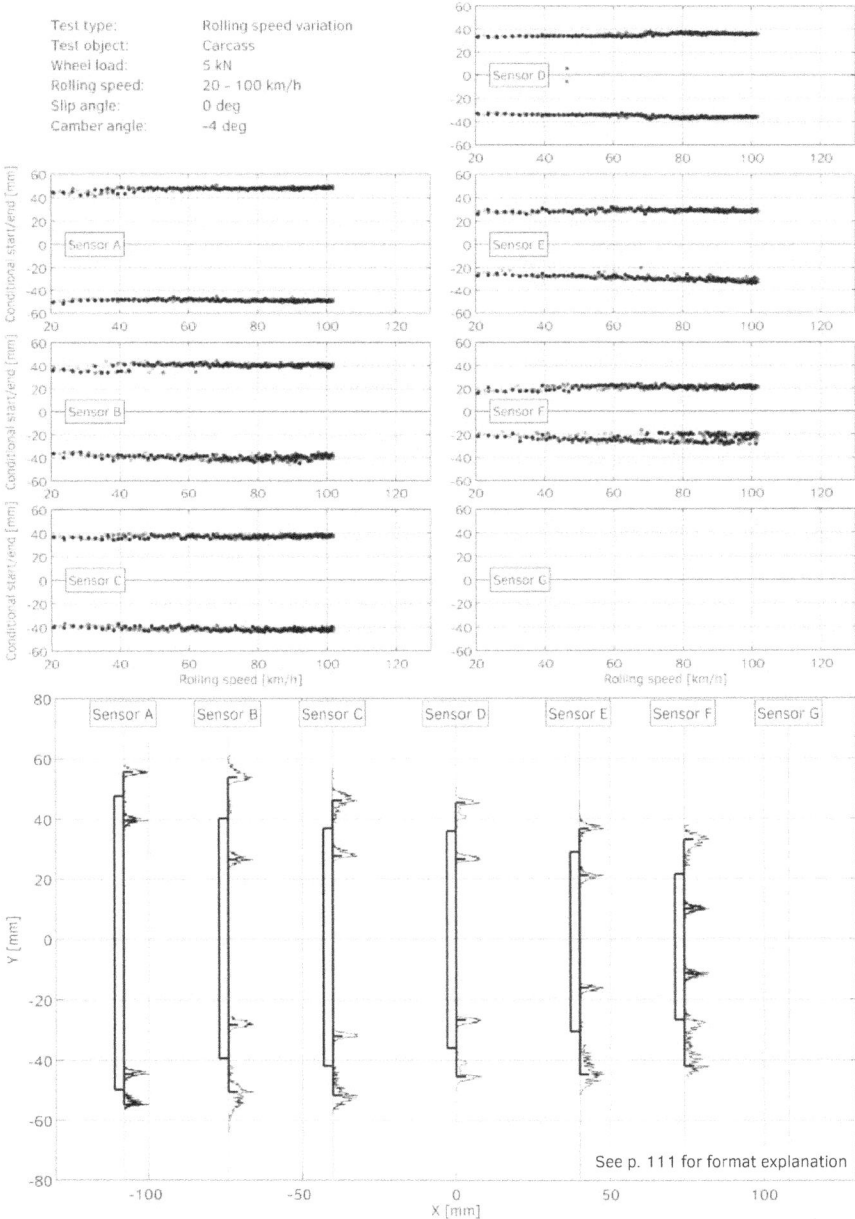

See p. 111 for format explanation

A.13. Rolling speed variation: Carcass, 7 kN, camber 0 deg

Test type: Rolling speed variation
Test object: Carcass
Wheel load: 7 kN
Rolling speed: 20 - 100 km/h
Slip angle: 0 deg
Camber angle: 0 deg

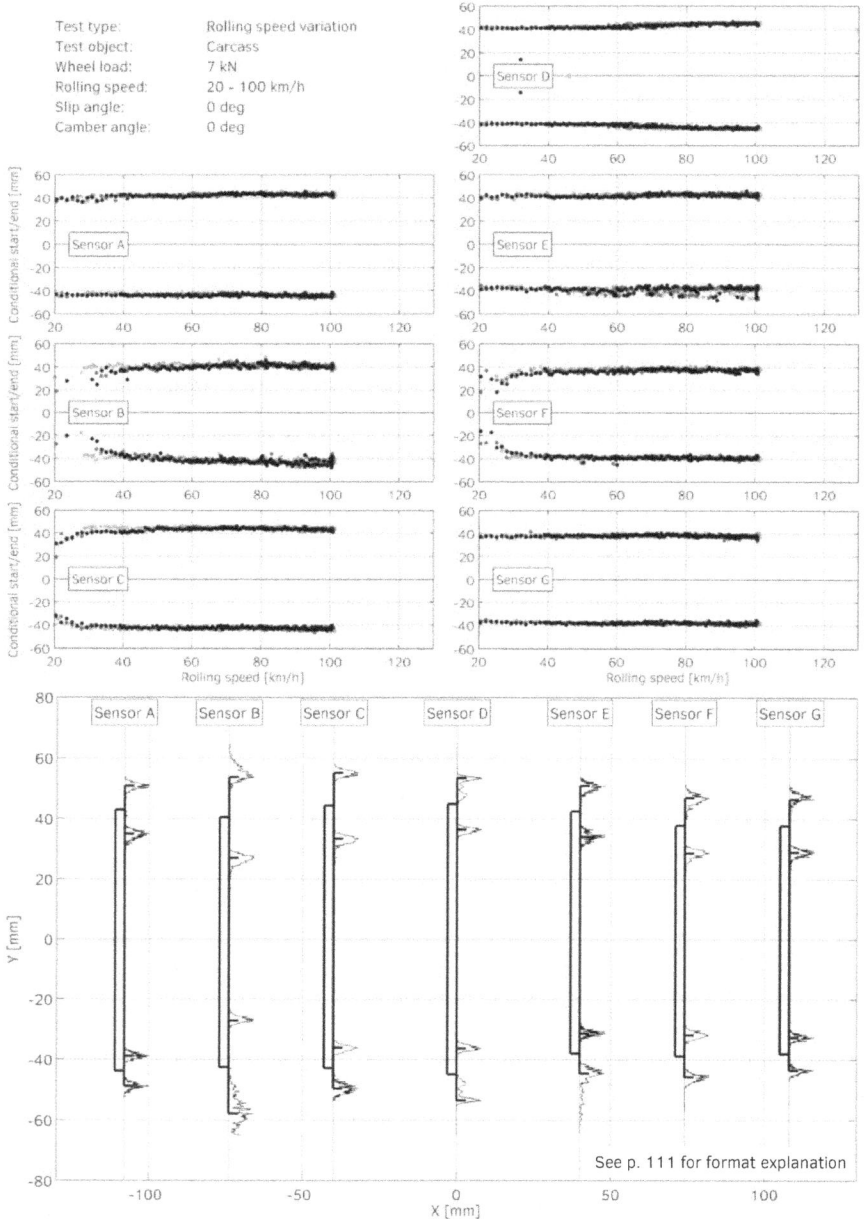

See p. 111 for format explanation

A.14. Rolling speed variation: Carcass, 7 kN, camber −4 deg

Test type:	Rolling speed variation
Test object:	Carcass
Wheel load:	7 kN
Rolling speed:	20 - 100 km/h
Slip angle:	0 deg
Camber angle:	−4 deg

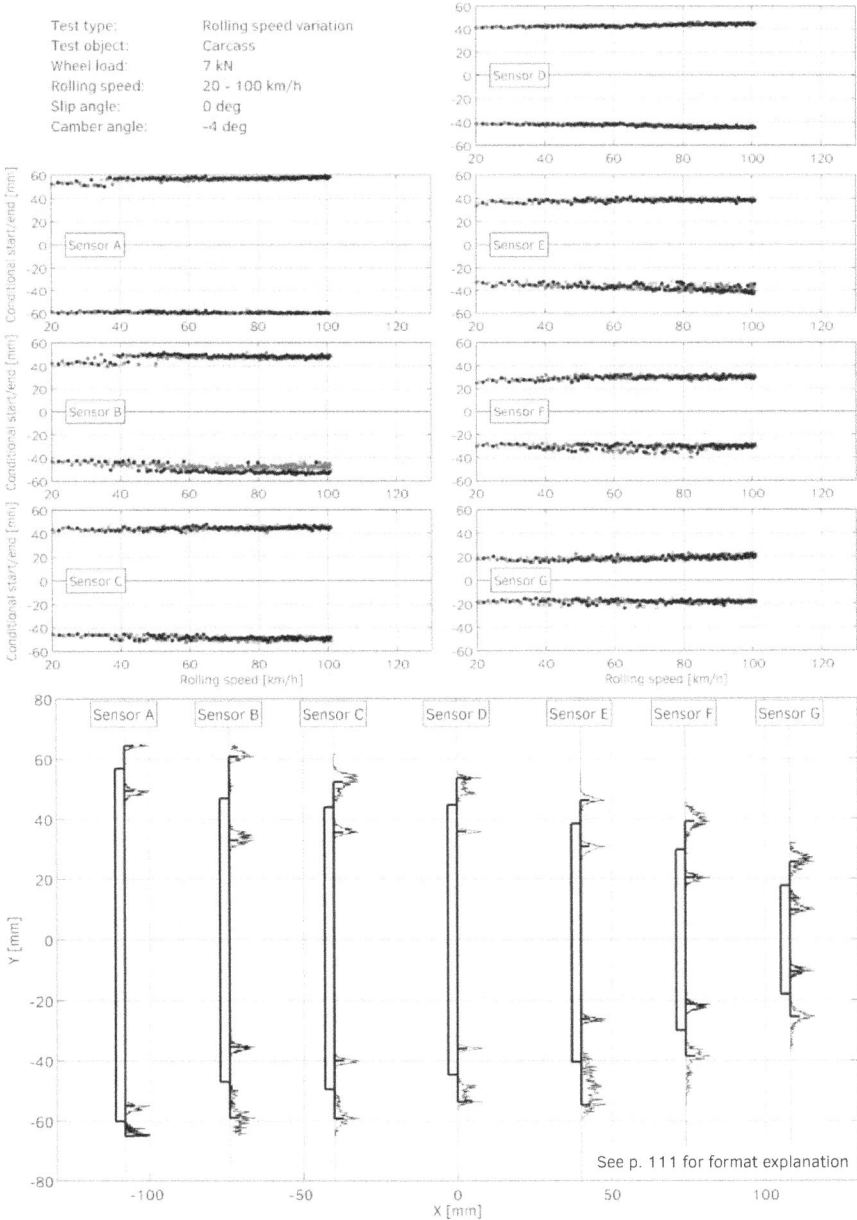

See p. 111 for format explanation

A.15. Rolling speed variation of cornering tire: 3 kN

Test type:	Rolling speed variation
Test object:	Tire
Wheel load:	3 kN
Rolling speed:	20 - 120 km/h
Slip angle:	3 deg
Camber angle:	0 deg

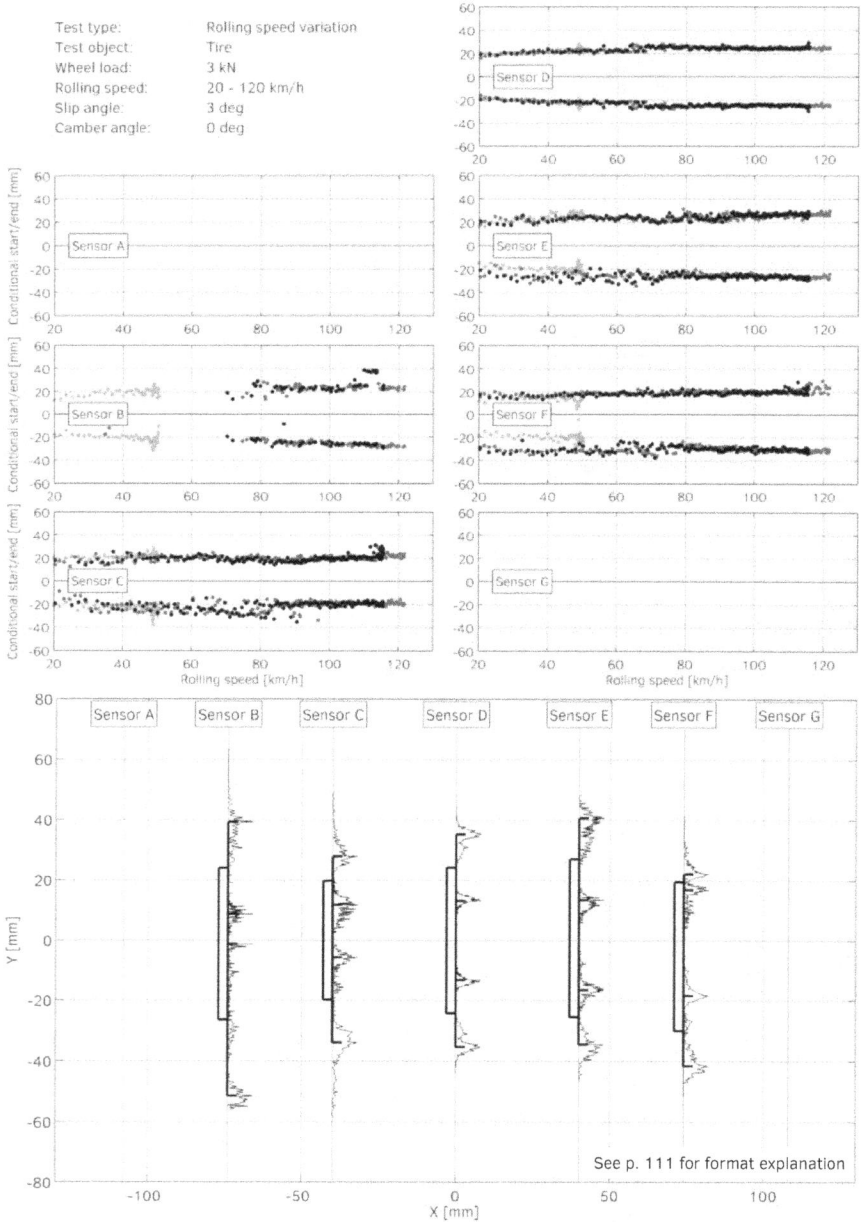

See p. 111 for format explanation

A.16. Rolling speed variation of cornering tire: 5 kN

Test type:	Rolling speed variation
Test object:	Tire
Wheel load:	5 kN
Rolling speed:	20 - 100 km/h
Slip angle:	3 deg
Camber angle:	0 deg

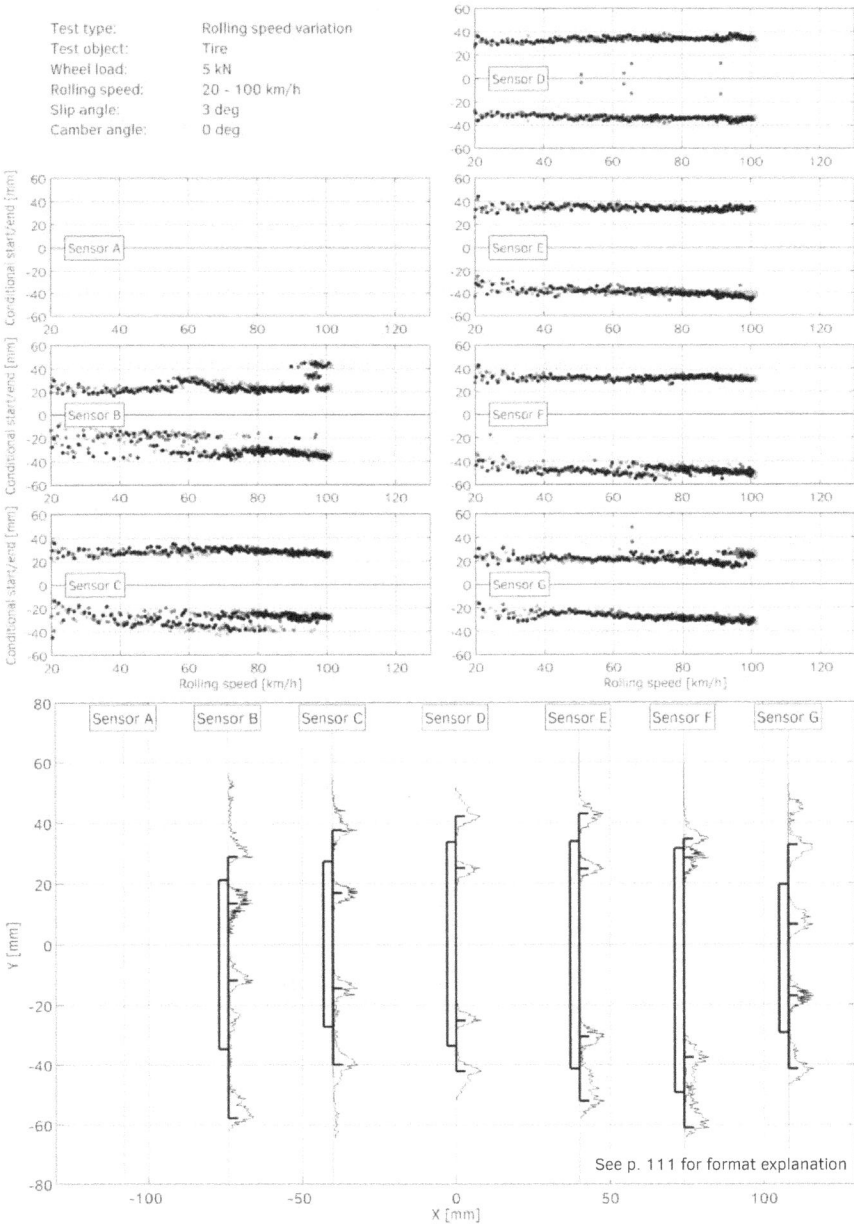

See p. 111 for format explanation

A.17. Rolling speed variation of cornering tire: 7 kN

Test type:	Rolling speed variation
Test object:	Tire
Wheel load:	7 kN
Rolling speed:	20 - 100 km/h
Slip angle:	3 deg
Camber angle:	0 deg

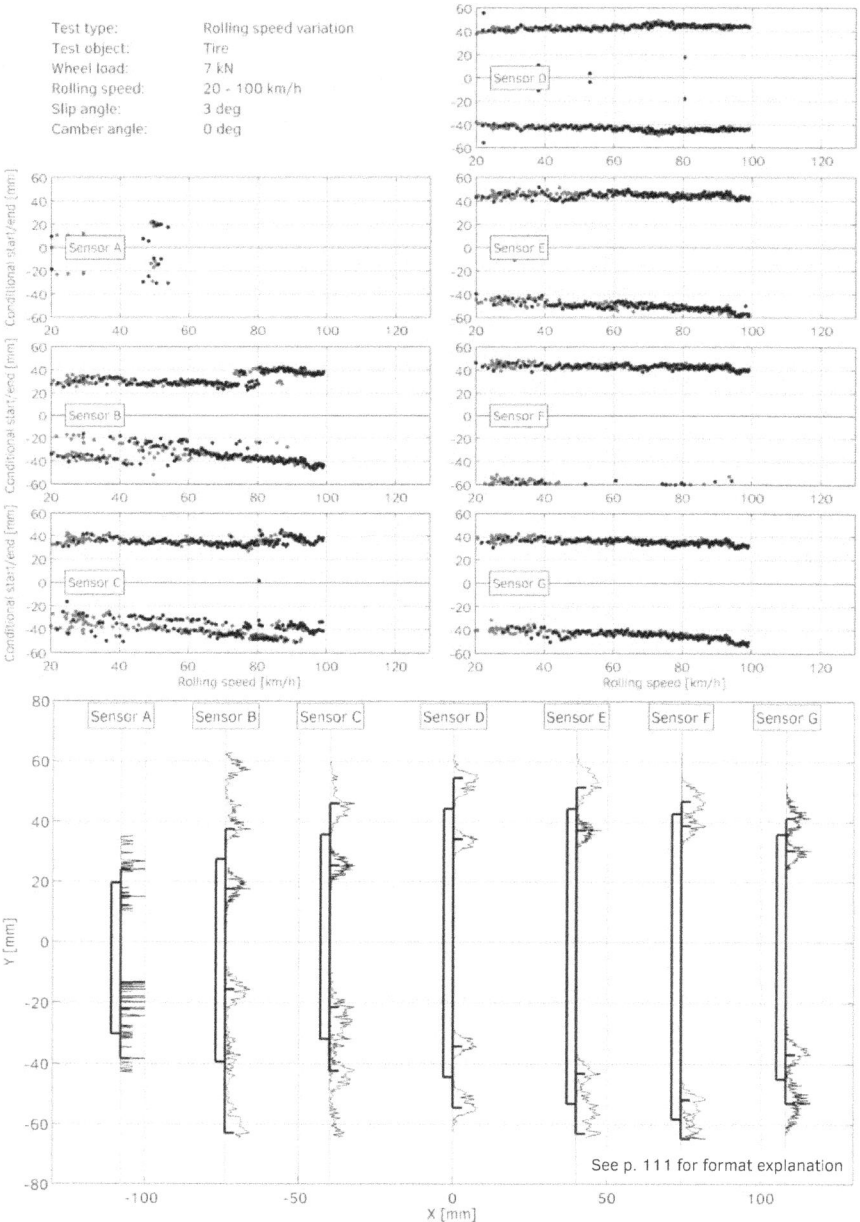

See p. 111 for format explanation

A.18. Rolling speed variation of cornering carcass: 7 kN

Test type: Rolling speed variation
Test object: Carcass
Wheel load: 7 kN
Rolling speed: 20 - 110 km/h
Slip angle: 1 deg
Camber angle: 0 deg

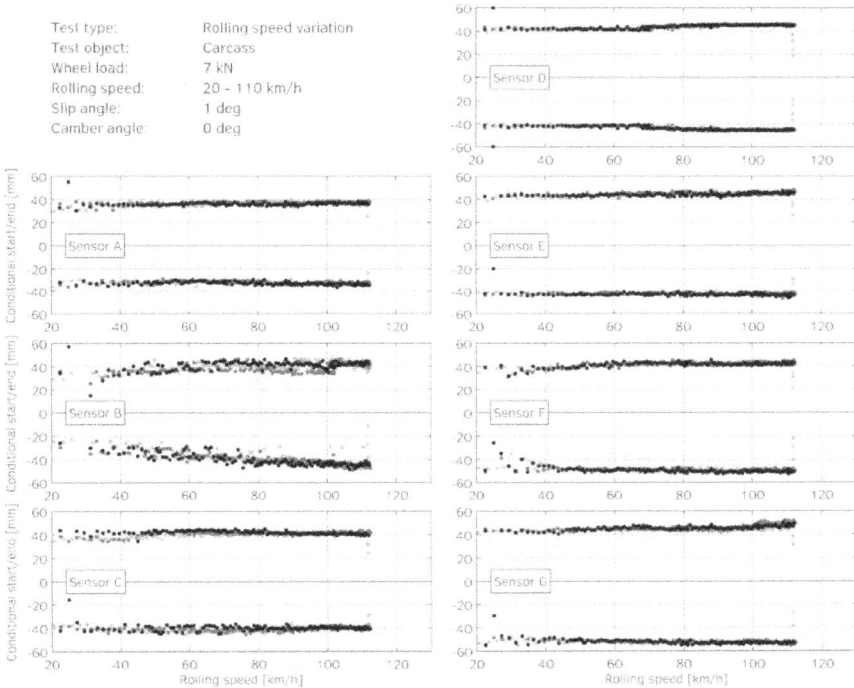

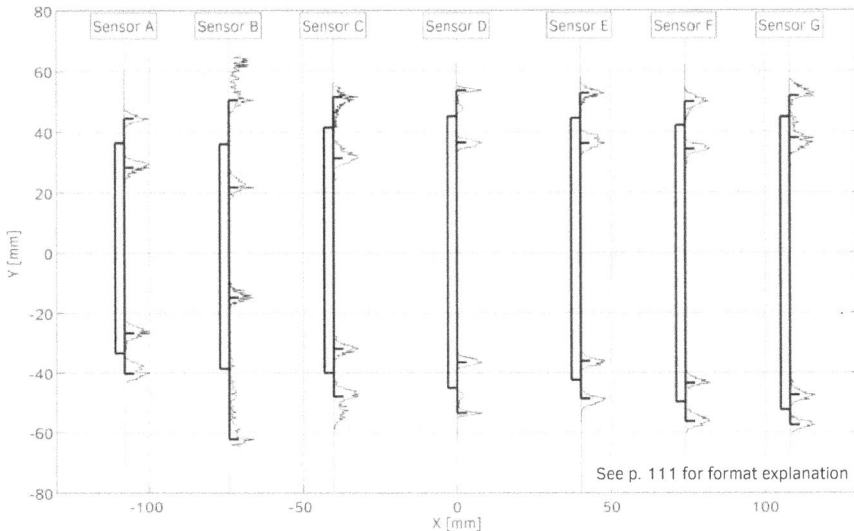

See p. 111 for format explanation

A.19. Camber variation: Tire, 3-5-7 kN, 60 km/h

Test type: Camber angle variation
Test object: Tire
Wheel load: 3 – 5 – 7 kN
Rolling speed: 60 km/h
Slip angle: 0 deg
Camber angle: -6 – +6 deg

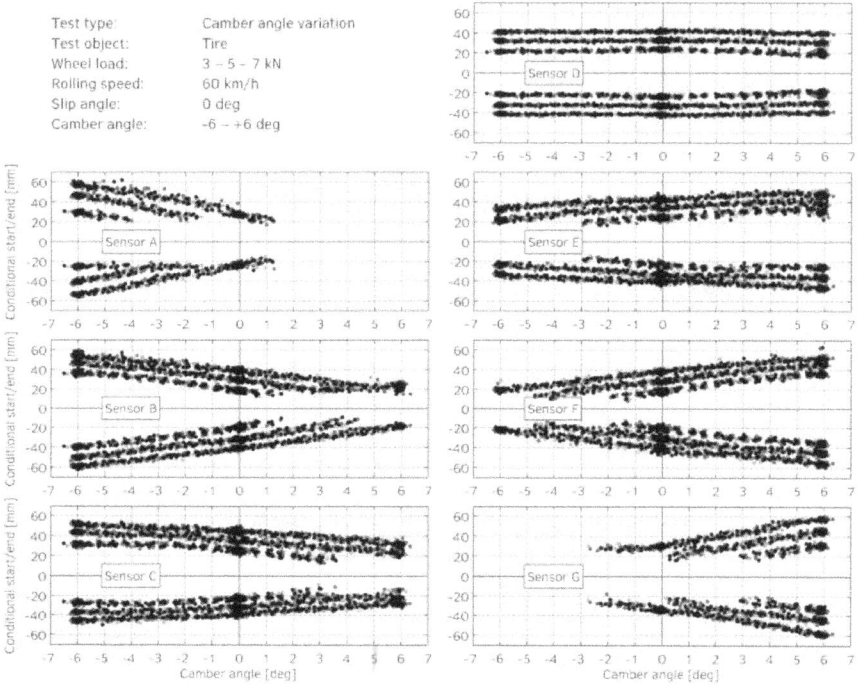

Interpolated leading and trailing edges for 7 kN wheel load:

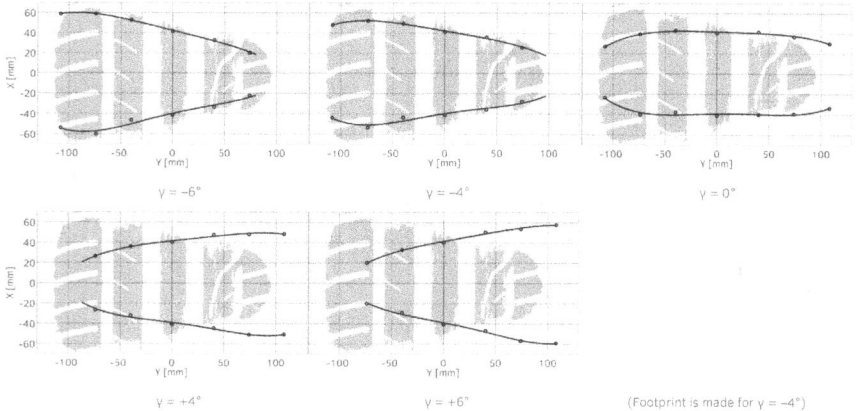

$\gamma = -6°$ $\gamma = -4°$ $\gamma = 0°$

$\gamma = +4°$ $\gamma = +6°$ (Footprint is made for $\gamma = -4°$)

A.20. Camber variation: Carcass, 3-5-7 kN, 60 km/h

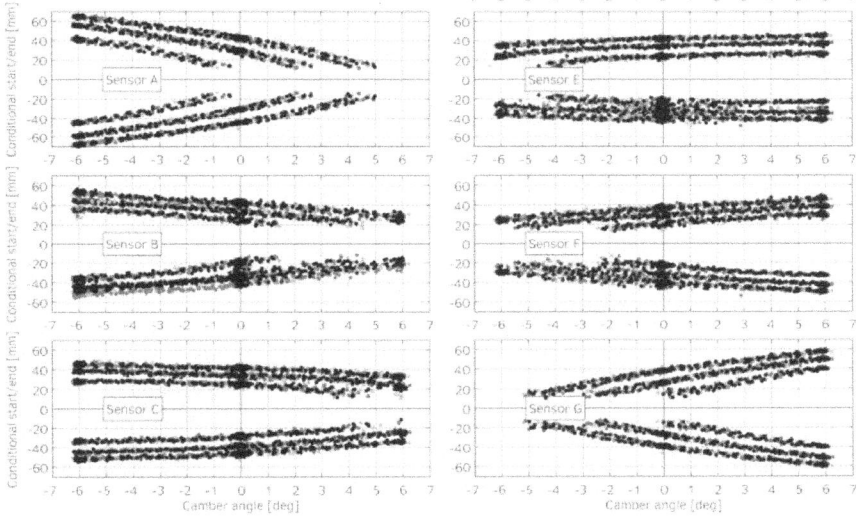

Test type: Camber angle variation
Test object: Carcass
Wheel load: 3 – 5 – 7 kN
Rolling speed: 60 km/h
Slip angle: 0 deg
Camber angle: -6 – +6 deg

A.21. Slip angle variation: Tire, 3-5-7 kN, 60 km/h, ±3 deg

Test type:	Slip angle variation (linear range)
Test object:	Tire
Wheel load:	3 – 5 – 7 kN
Rolling speed:	60 km/h
Slip angle:	-3 – +3 deg
Camber angle:	0 deg

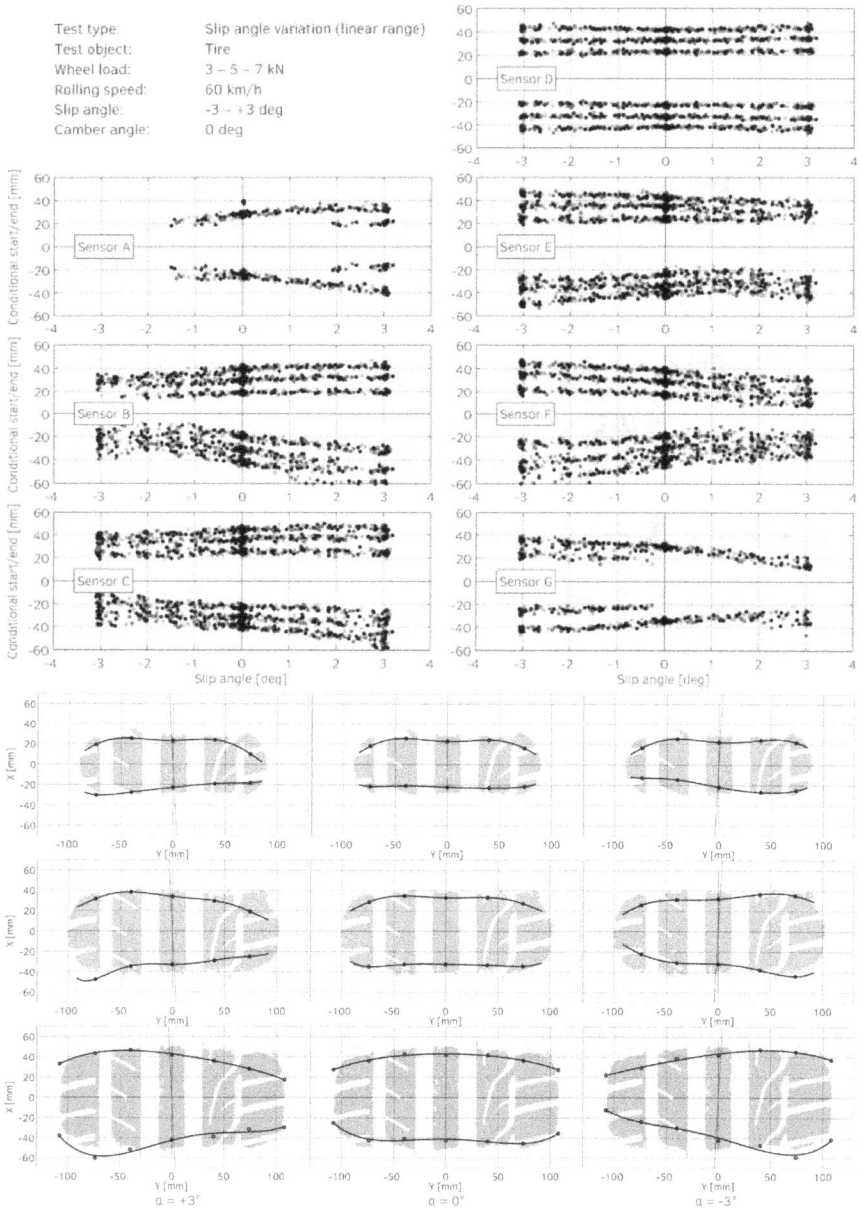

A.22. Slip angle variation: Carcass, 3-5-7 kN, 60 km/h, ±3 deg

Test type: Slip angle variation (linear range)
Test object: Carcass
Wheel load: 3 – 5 – 7 kN
Rolling speed: 60 km/h
Slip angle: -3 – +3 deg
Camber angle: 0 deg

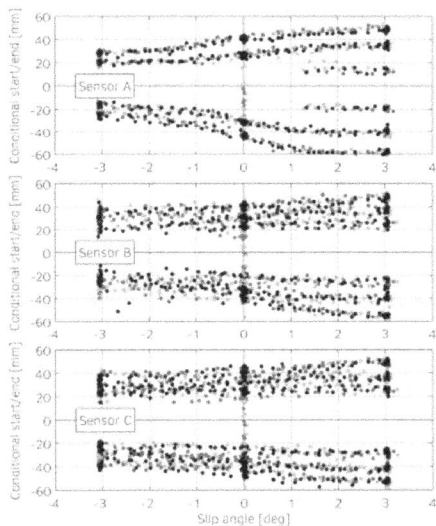

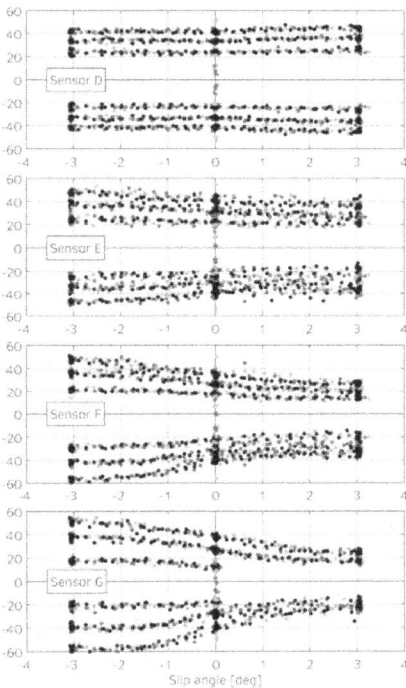

A.23. Slip angle variation: Carcass, 3-5-7 kN, 60 km/h, ±1 deg

Test type: Slip angle variation (linear range)
Test object: Carcass
Wheel load: 3 – 5 – 7 kN
Rolling speed: 60 km/h
Slip angle: -1 – +1 deg
Camber angle: 0 deg

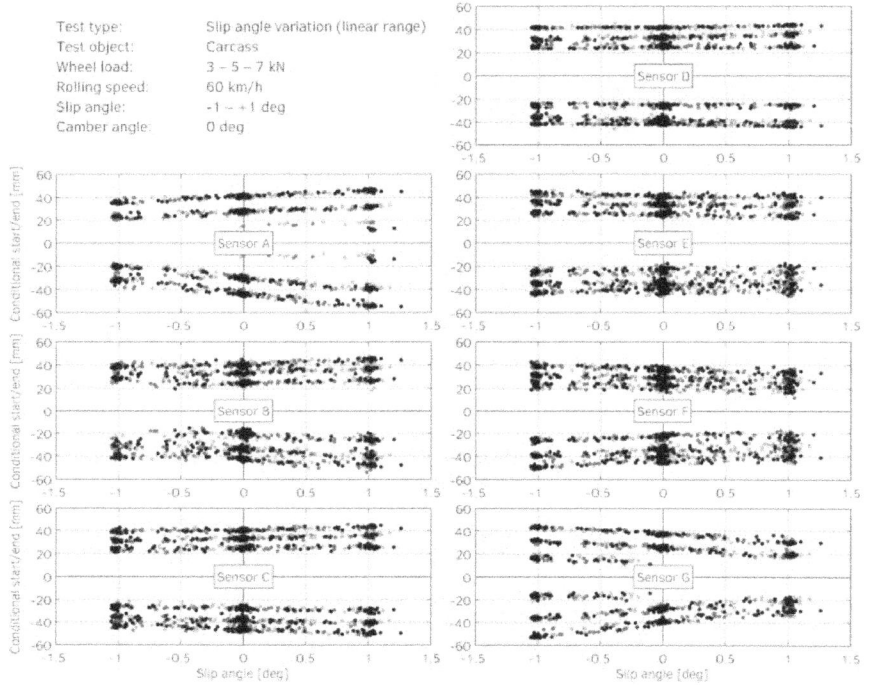

A.24. Slip angle variation: Tire, 5 kN, 60 km/h, −12 ... +1 deg

Test type:	Slip angle variation (nonlinear range)
Test object:	Tire
Wheel load:	5 kN
Rolling speed:	60 km/h
Slip angle:	-12 − +1 deg
Camber angle:	0 deg

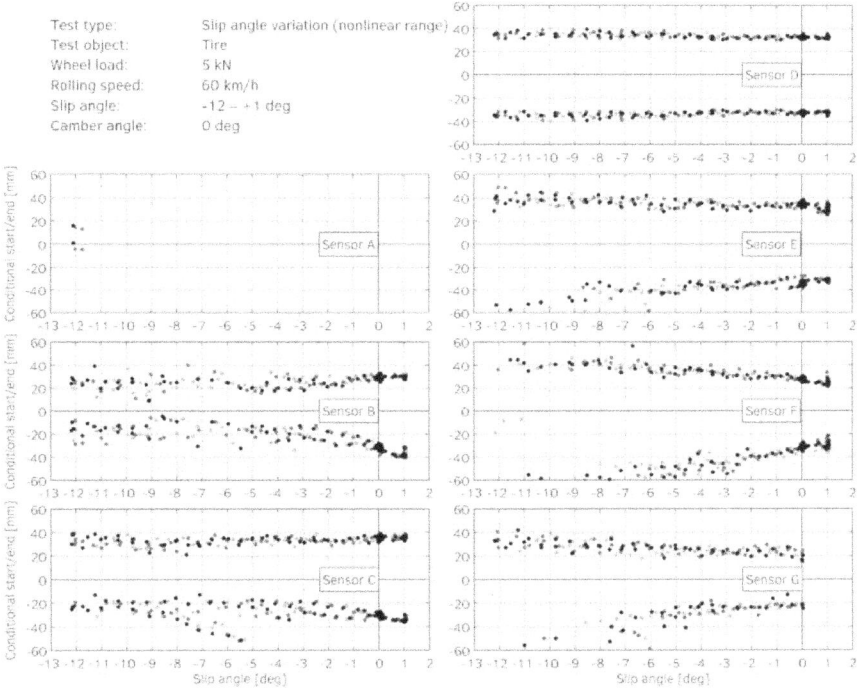

A.25. Slip angle variation: Tire, 5 kN, 60 km/h, −1 ... +12 deg

Test type: Slip angle variation (nonlinear range)
Test object: Tire
Wheel load: 5 kN
Rolling speed: 60 km/h
Slip angle: −1 – +12 deg
Camber angle: 0 deg

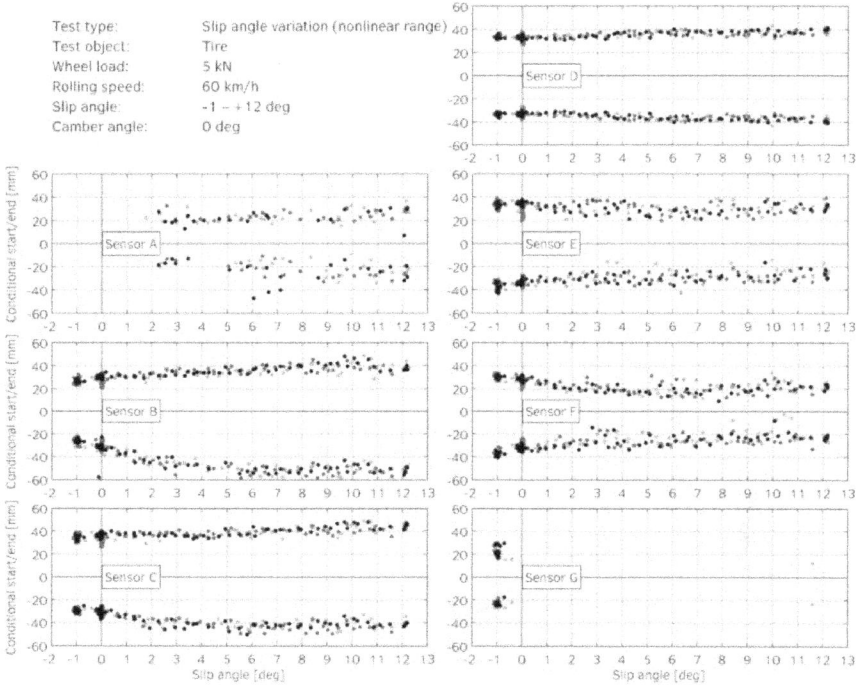

A.26. Slip angle variation: Tire, 5 kN, 90 km/h, −1 ... +12 deg

Test type: Slip angle variation (nonlinear range)
Test object: Tire
Wheel load: 5 kN
Rolling speed: 90 km/h
Slip angle: −1 ... +12 deg
Camber angle: 0 deg

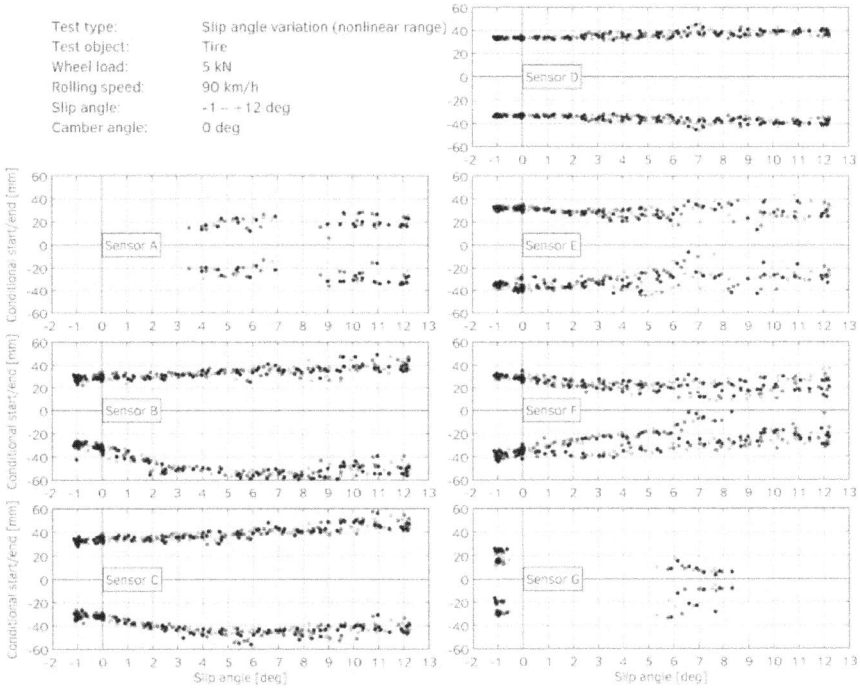

A.27. Slip angle variation: Tire, 7 kN, 60 km/h, −1 ... +12 deg

Test type:	Slip angle variation (nonlinear range)
Test object:	Tire
Wheel load:	7 kN
Rolling speed:	60 km/h
Slip angle:	-1 - +12 deg
Camber angle:	0 deg

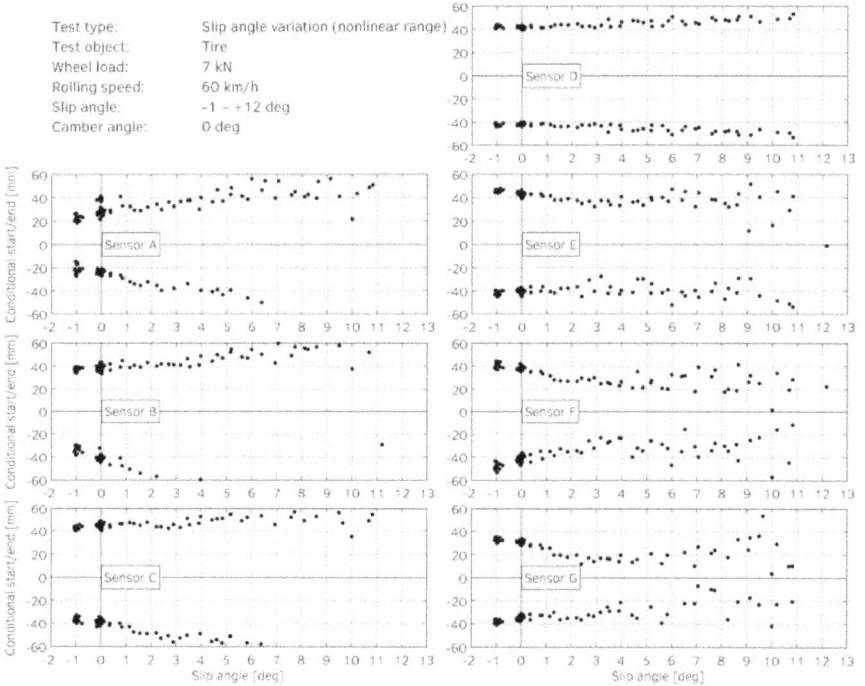

A.28. Slip angle variation: Tire, 5 kN, 60 km/h, −12 … +1 deg

Test type:	Slip angle variation (nonlinear range)
Test object:	Tire
Wheel load:	5 kN
Rolling speed:	60 km/h
Slip angle:	-12 – +1 deg
Camber angle:	0 deg

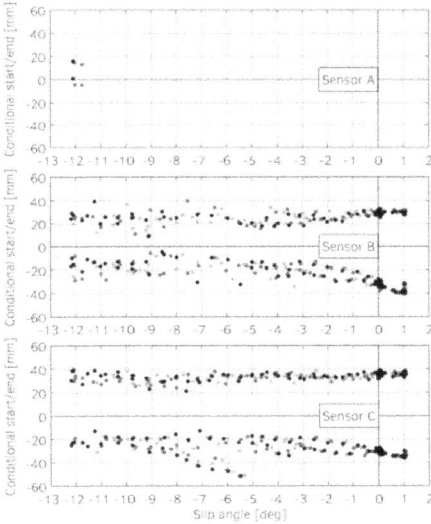

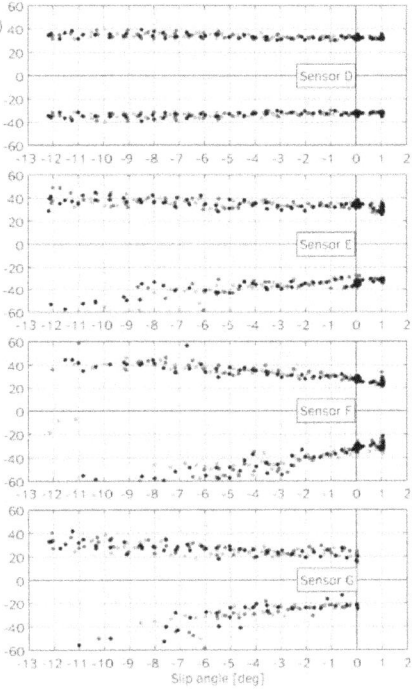

A.29. Slip angle variation: Tire, 5 kN, 60 km/h, −12 … +1 deg, cambered

Test type: Slip angle variation (nonlinear range)
Test object: Tire
Wheel load: 5 kN
Rolling speed: 60 km/h
Slip angle: −12 – +1 deg
Camber angle: −4 deg

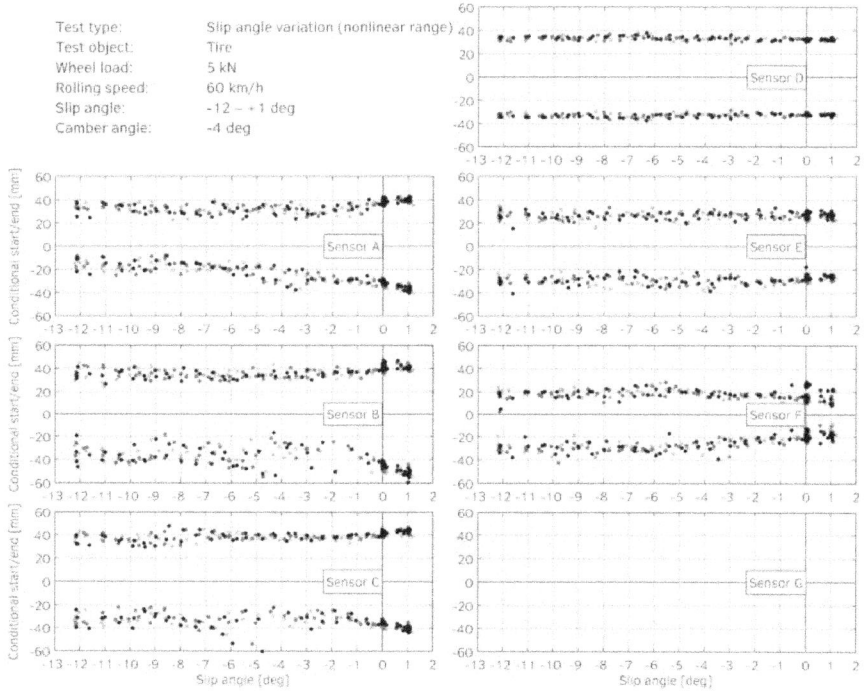

A.30. Optical measurement: Carcass lateral stiffness (eight cameras)

Wheel load: 3 kN

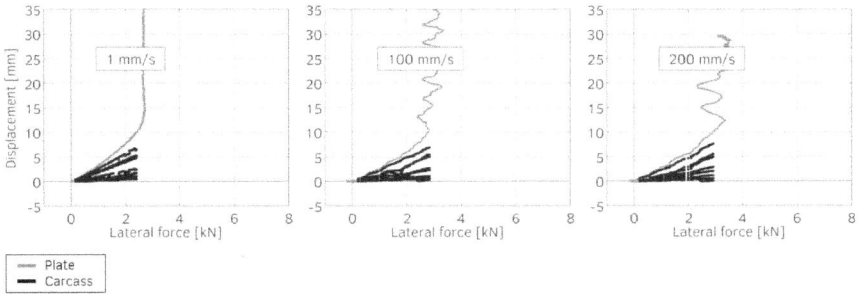

Wheel load: 5 kN

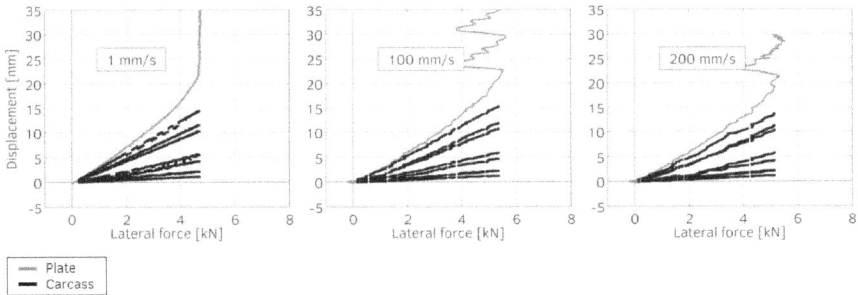

Wheel load: 7 kN

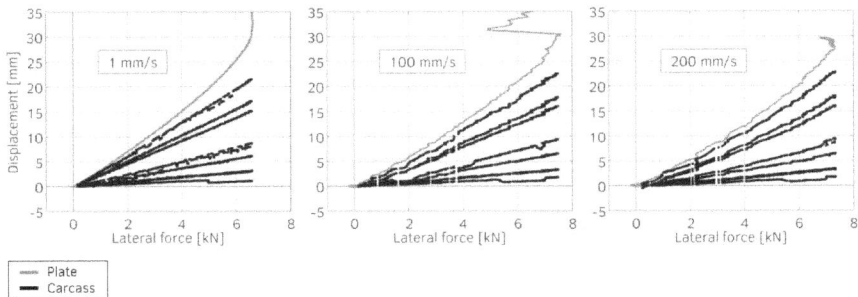

A.31. Optical measurement: Carcass lateral stiffness (main camera)

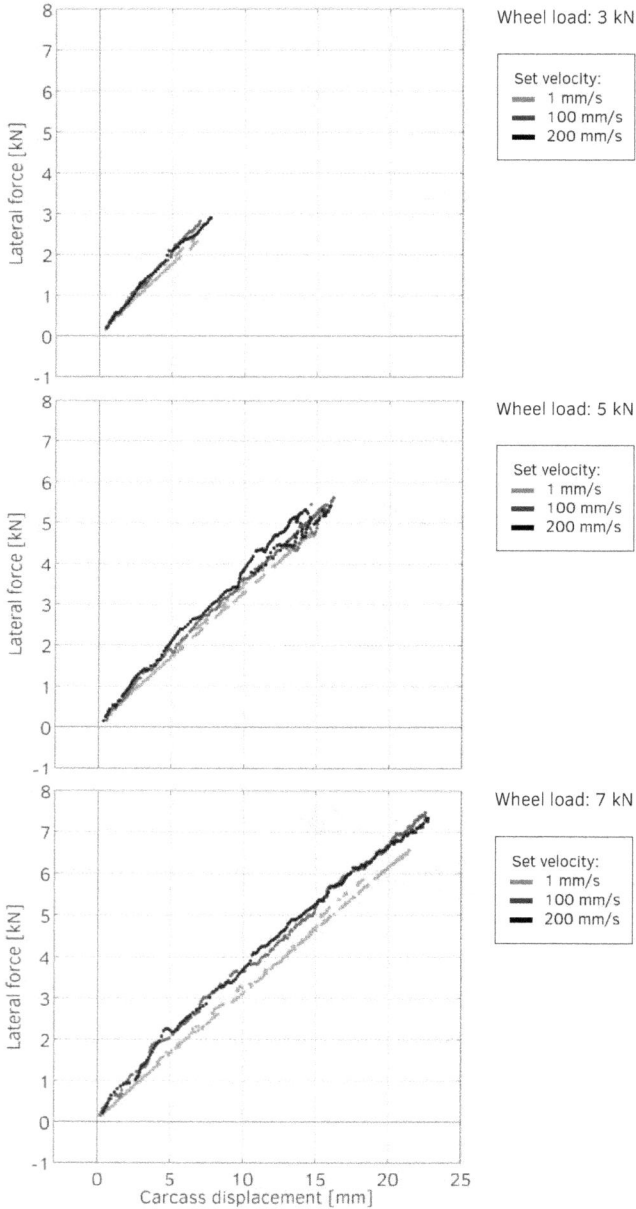

Wheel load: 3 kN

Set velocity:
- 1 mm/s
- 100 mm/s
- 200 mm/s

Wheel load: 5 kN

Set velocity:
- 1 mm/s
- 100 mm/s
- 200 mm/s

Wheel load: 7 kN

Set velocity:
- 1 mm/s
- 100 mm/s
- 200 mm/s

A.32. Optical measurement: Carcass lateral stiffness (separated)

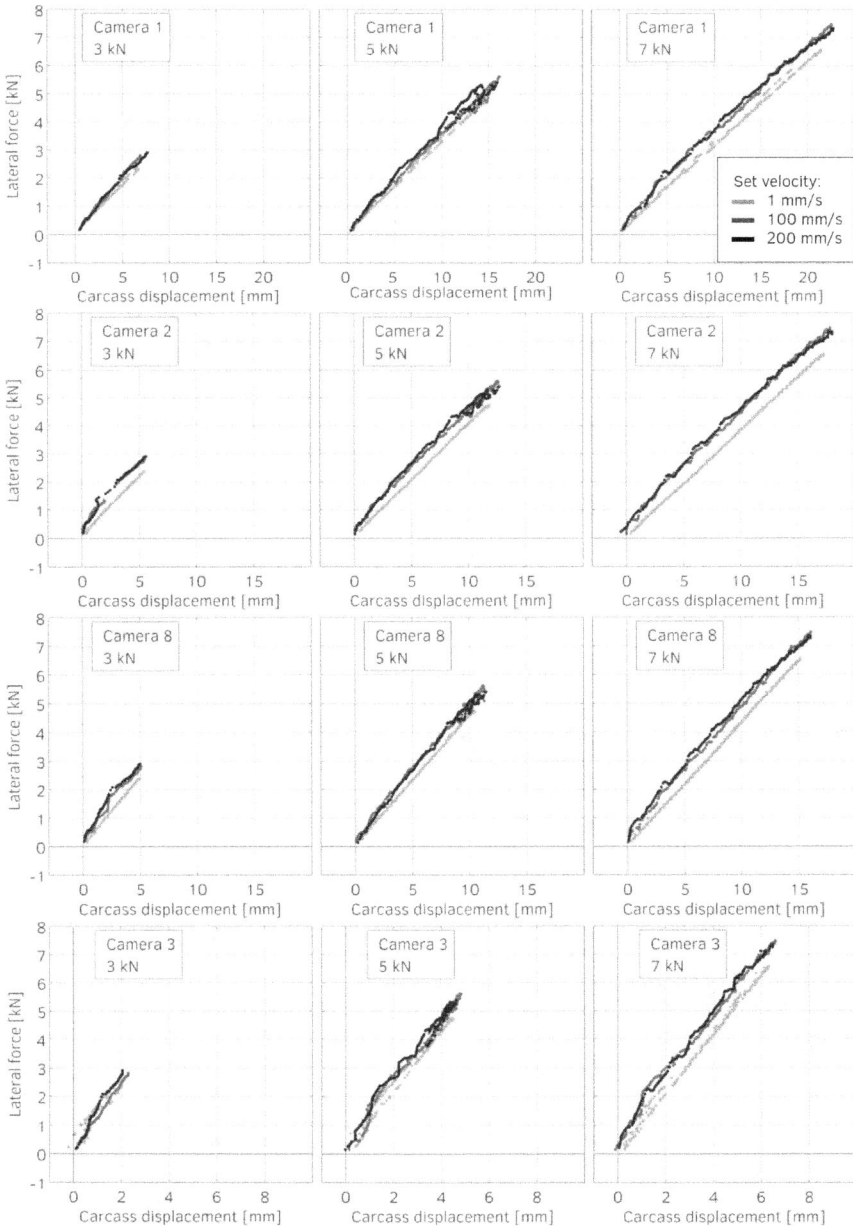

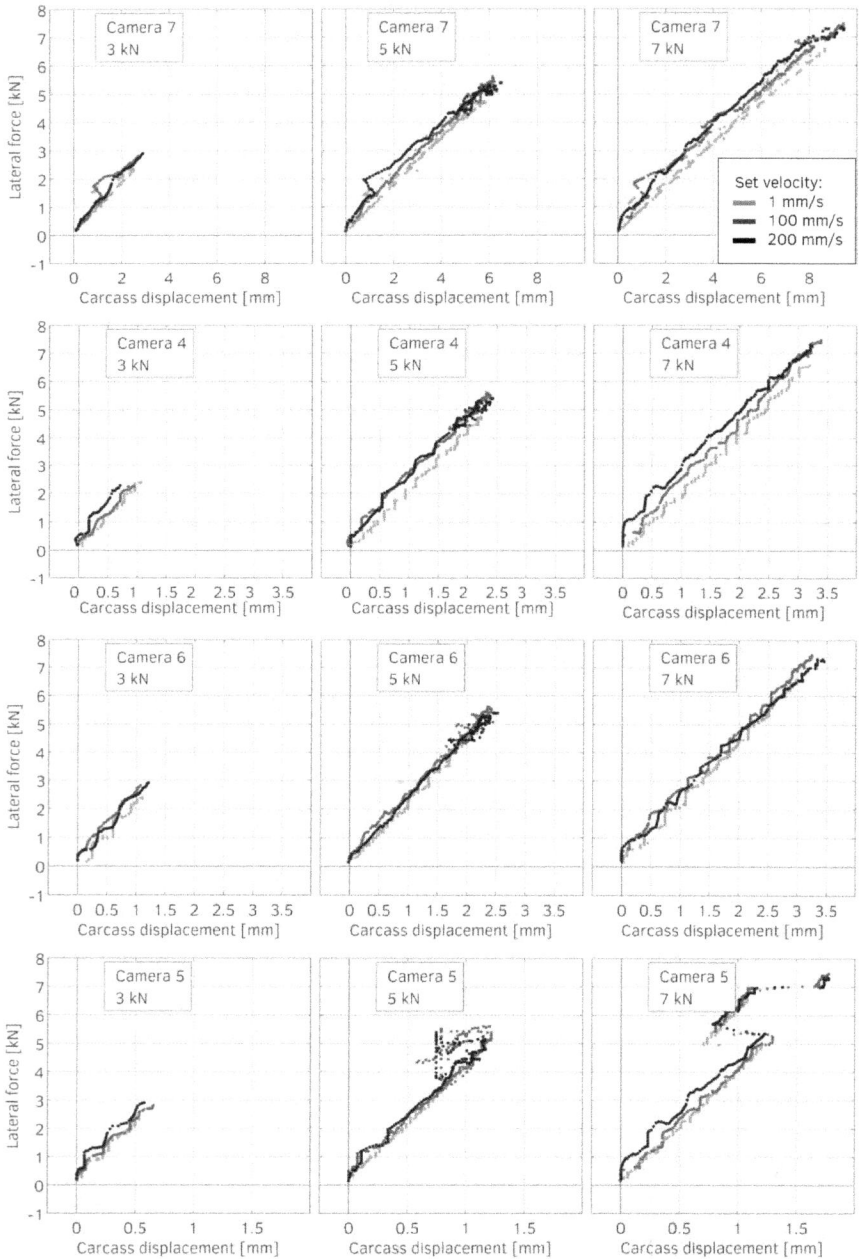

A.33. Stereocamera-based measurement: cornering stiffness, 5 kN

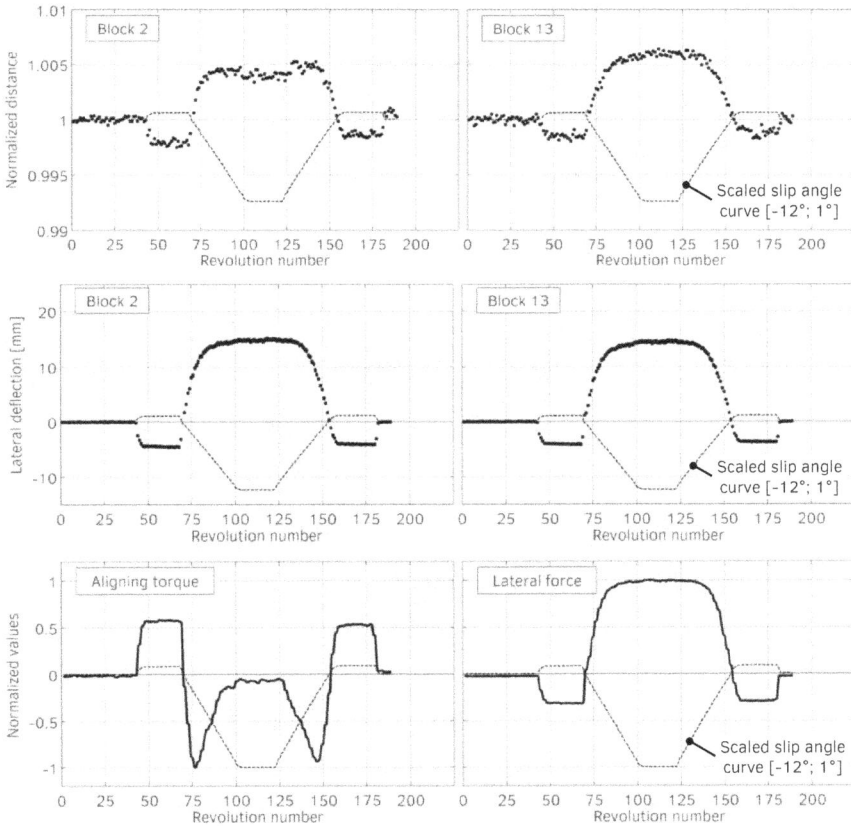

A.34. Stereocamera-based measurement: cornering stiffness, 7 kN

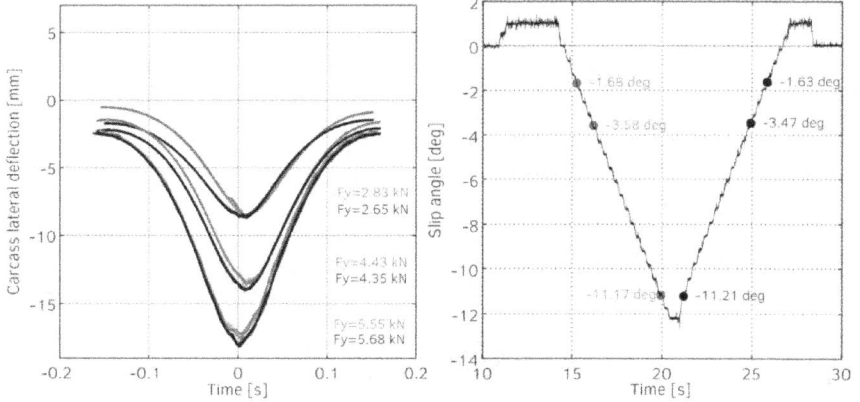

Curves in the linear range, measured with increasing and decreasing slip angle, are different. Curves outside linear range, measured with increasing and decreasing slip angle, become similar and finally same in case of full slide (e.g. 12 deg).

A.35. Model computing method issues

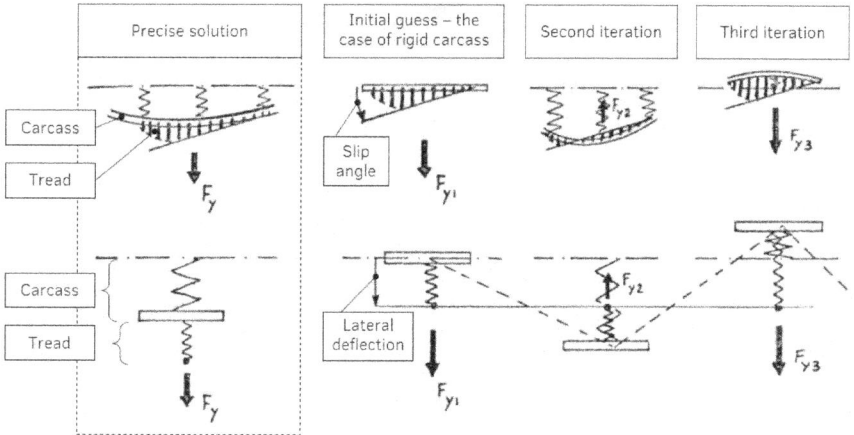

A.36. Validation with help of slip angle step (3 kN, 1-2-3 deg)

Maneuver: **slip angle step (excitation from 0° to set angle value and then back to 0°)**
Wheel load: **3 kN**
Slip angle: **1° (top), 2° (middle), 3° (bottom)**

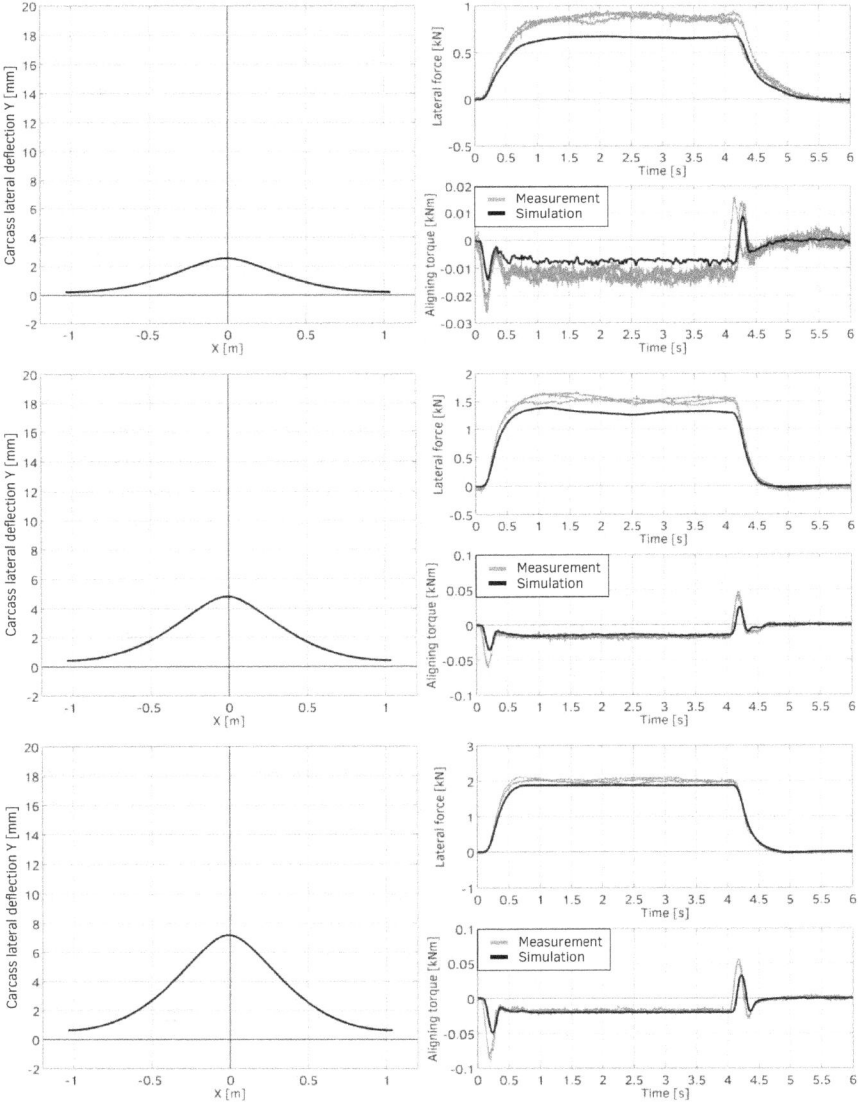

A.37. Validation with help of slip angle step (5 kN, 1-2-3 deg)

Maneuver: slip angle step (excitation from 0° to set angle value and then back to 0°)
Wheel load: 5 kN
Slip angle: 1° (top), 2° (middle), 3° (bottom)

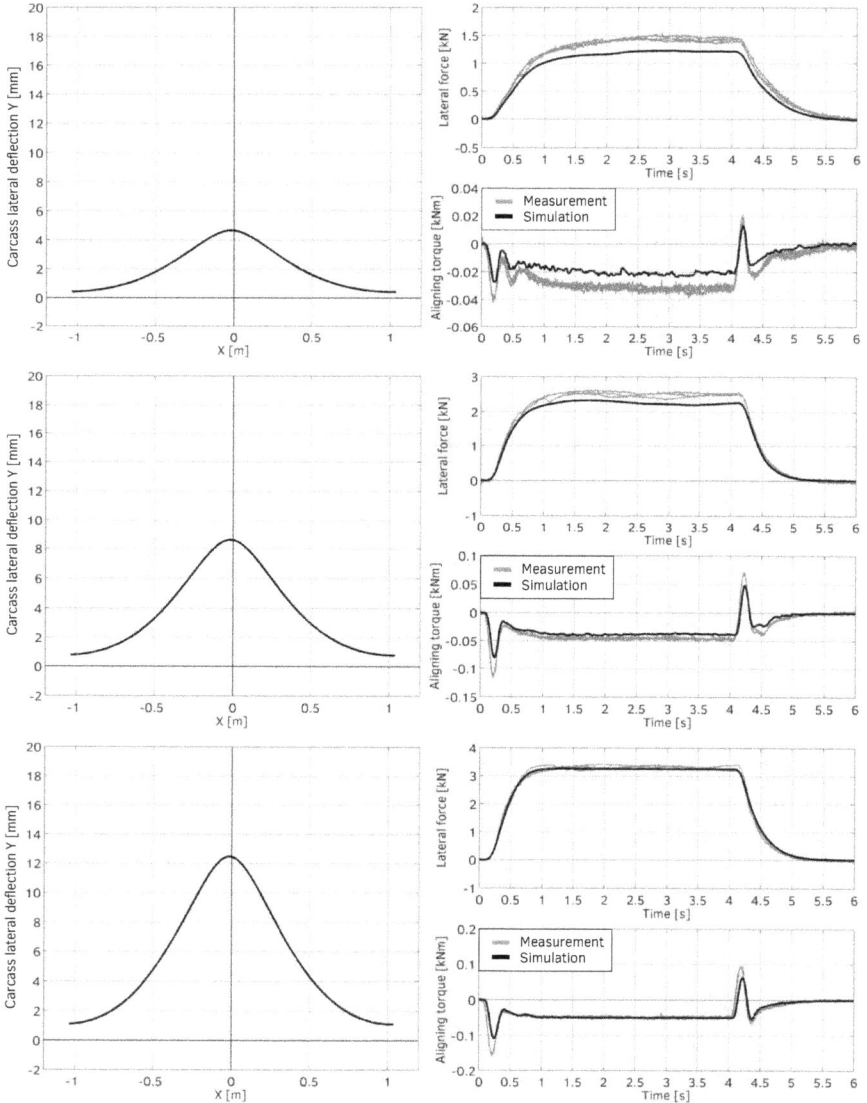

A.38. Validation with help of slip angle step (7 kN, 1-2-3 deg)

Maneuver: **slip angle step (excitation from 0° to set angle value and then back to 0°)**
Wheel load: **7 kN**
Slip angle: **1° (top), 2° (middle), 3° (bottom)**

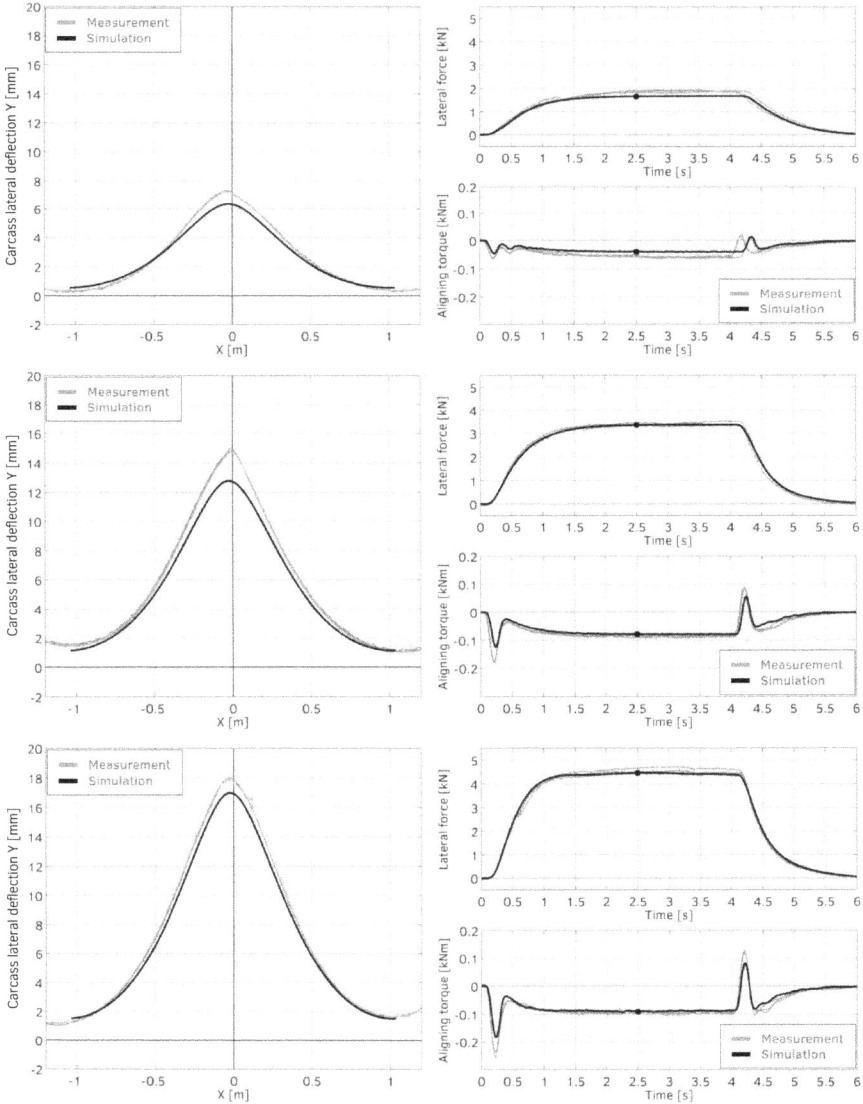

A.39. Validation with help of wheel load step

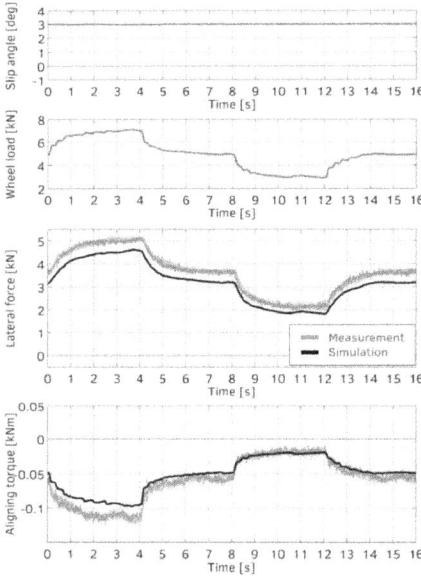

Validation with help of wheel load variation of 5-7-5-3-5 kN, slip angle is 3°.

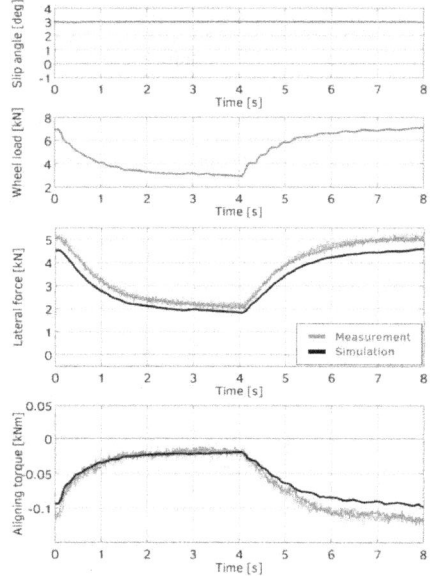

Validation with help of wheel load variation of 7-3-7 kN, slip angle is 3°.

A.40. Validation with help of combined transient excitation

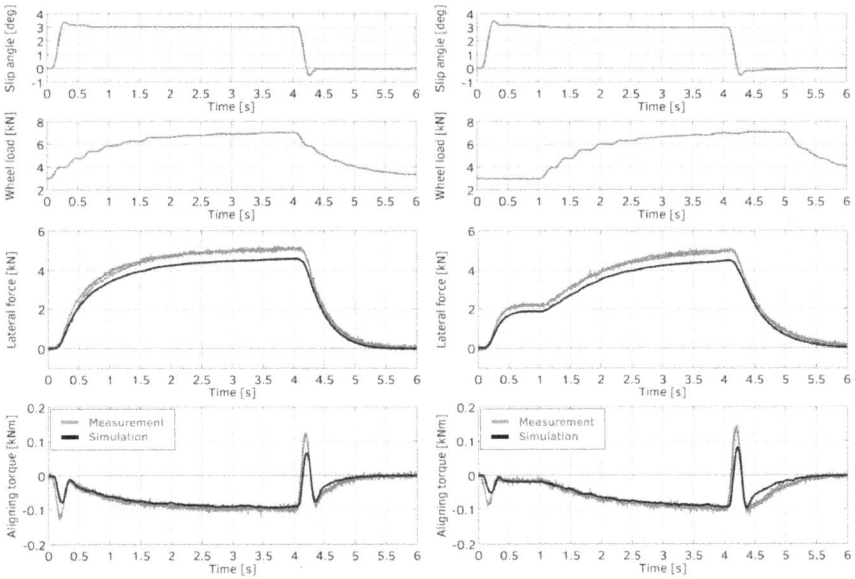

Validation with help of combined simultaneous excitation:

- Wheel load   3-7-3 kN
- Slip angle   0-3-0°

Validation with help of combined overlapping excitation:

- Wheel load   3-7-3 kN
- Slip angle   0-3-0°

A.41. Comparison of strain figures in turning phase and steady state

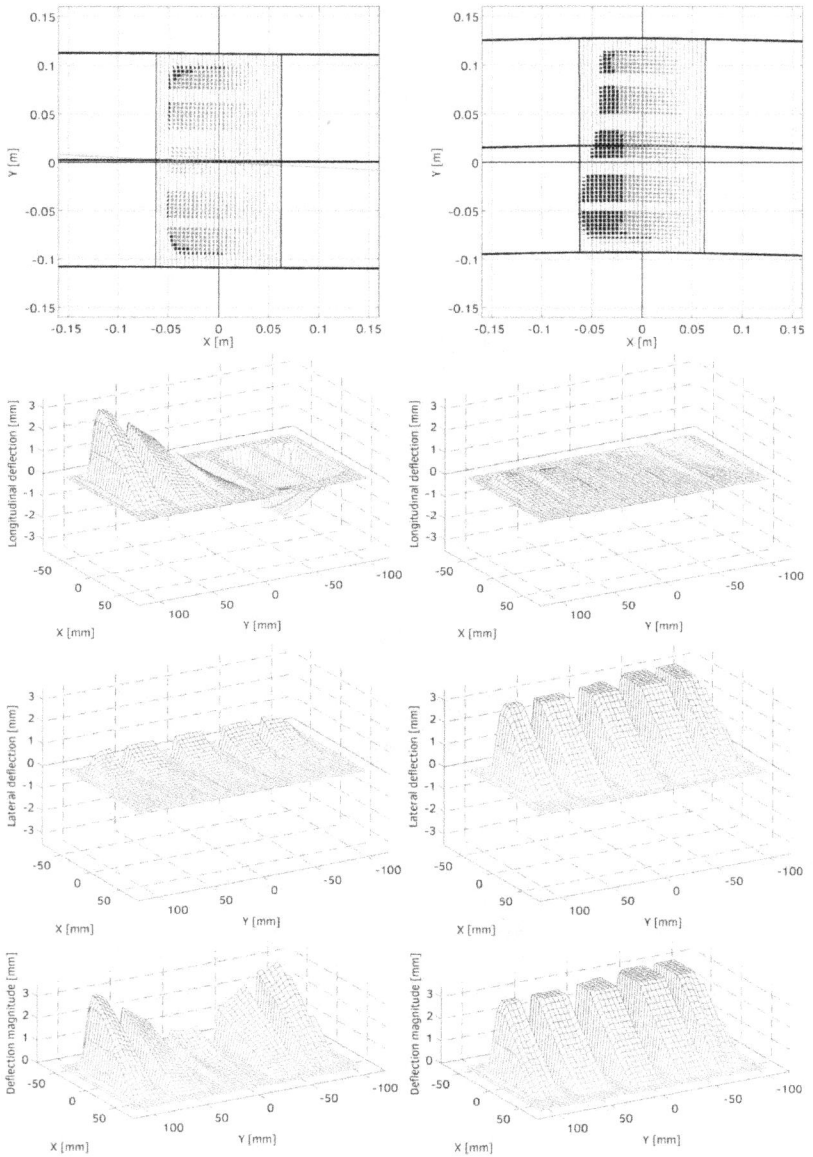

Strain figure and deflection distributions in the turning phase (Figure 3.20, 0.25 s).

Strain figure and deflection distributions in the steady state (Figure 3.20, 2 s).

A.42. Sensitivity analysis: Tread shear stiffness

Slip angle step maneuver (left diagram – deflected carcass, right diagrams – force and torque):

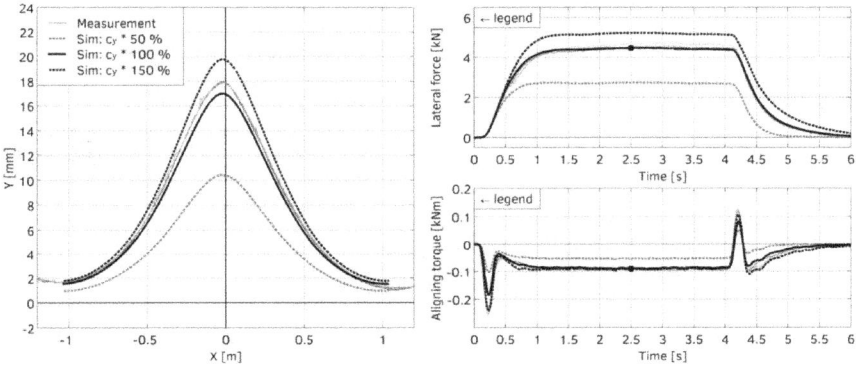

Lateral displacement maneuver (left diagram – deflected carcass, right diagram – lateral force):

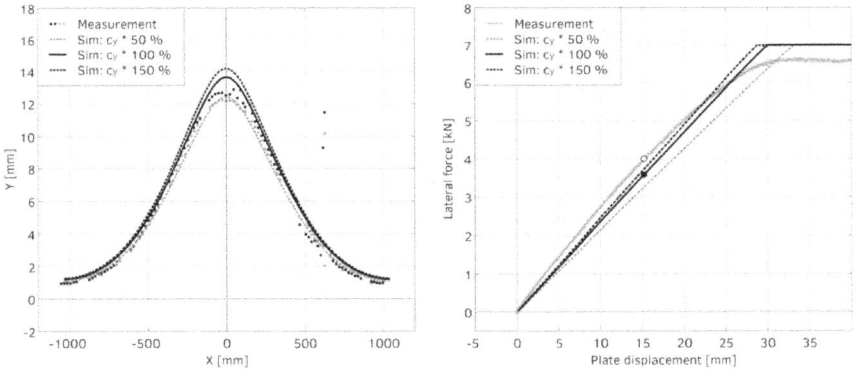

Torsional displacement maneuver (left diagram – deflected carcass, right diagram – bore torque):

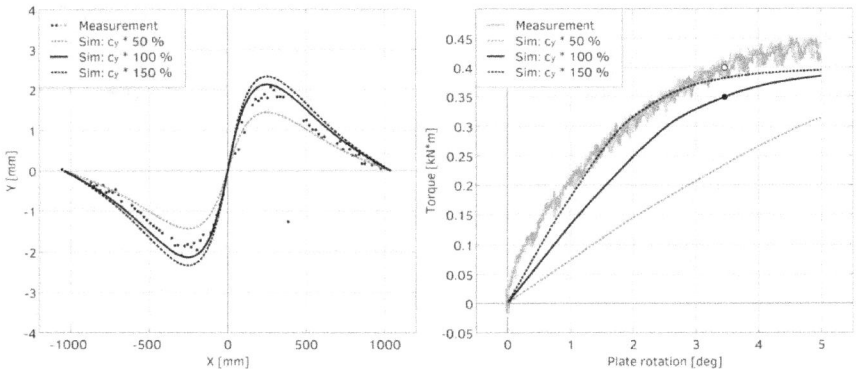

A.43. Sensitivity analysis: Carcass bending stiffness

Slip angle step maneuver (left diagram – deflected carcass, right diagrams – force and torque):

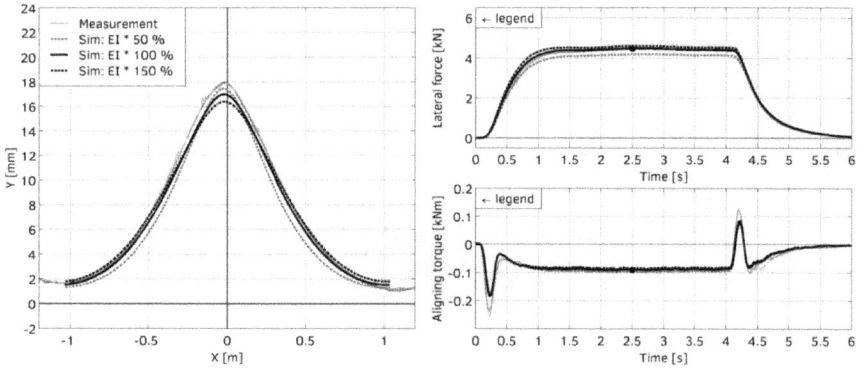

Lateral displacement maneuver (left diagram – deflected carcass, right diagram – lateral force):

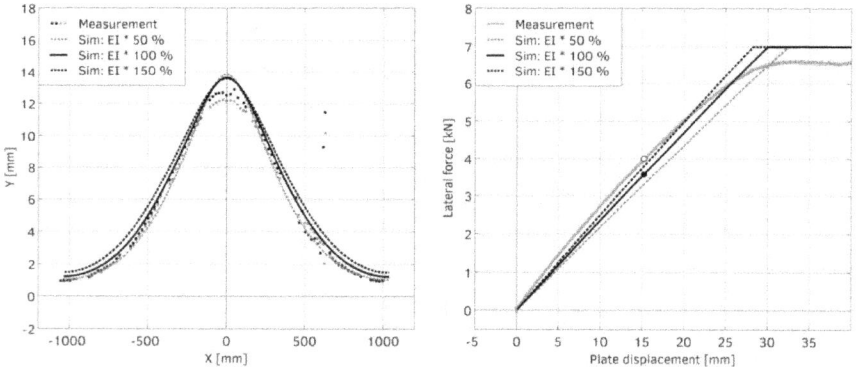

Torsional displacement maneuver (left diagram – deflected carcass, right diagram – bore torque):

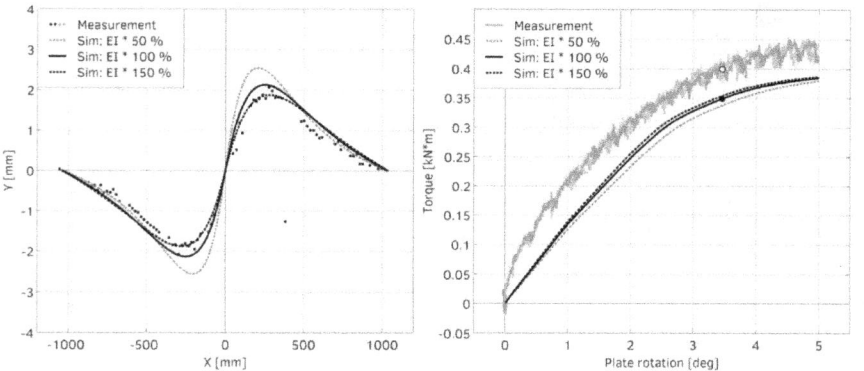

A.44. Sensitivity analysis: Carcass tensile force

Slip angle step maneuver (left diagram – deflected carcass, right diagrams – force and torque):

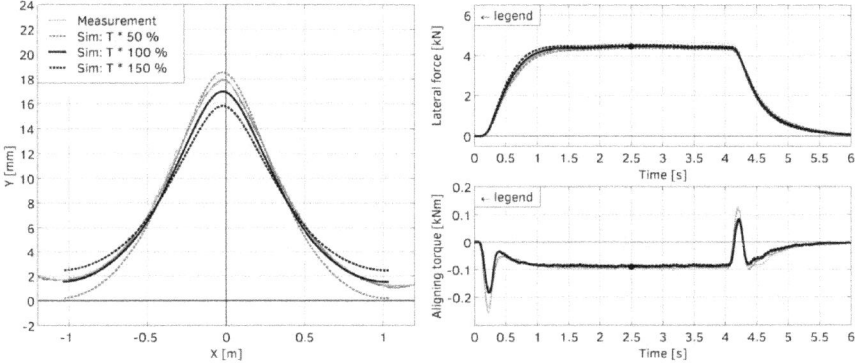

Lateral displacement maneuver (left diagram – deflected carcass, right diagram – lateral force):

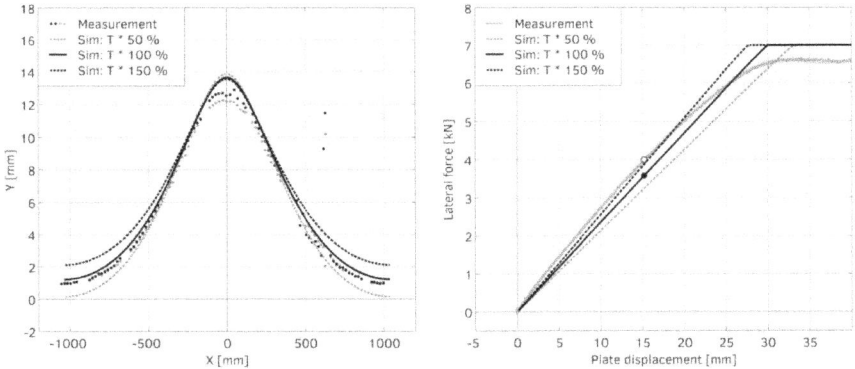

Torsional displacement maneuver (left diagram – deflected carcass, right diagram – bore torque):

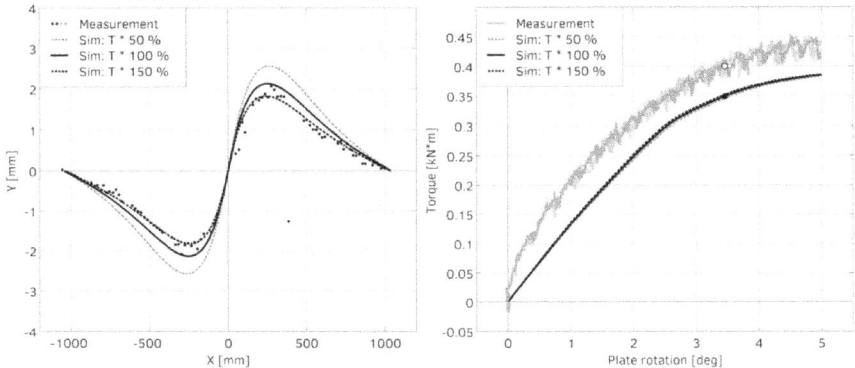

A.45. Sensitivity analysis: Carcass lateral flexibility

Slip angle step maneuver (left diagram – deflected carcass, right diagrams – force and torque):

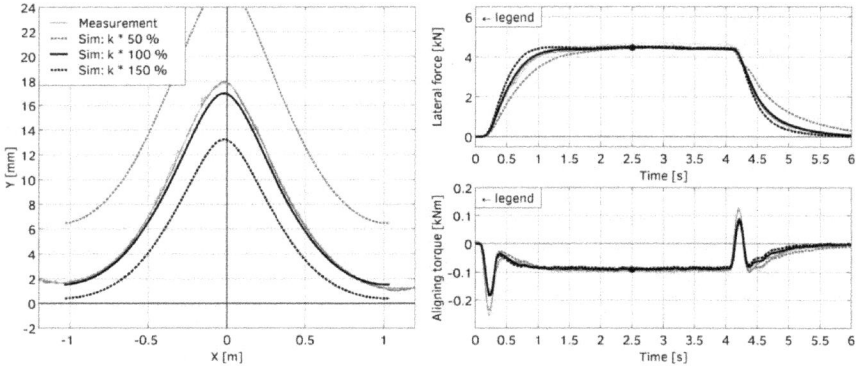

Lateral displacement maneuver (left diagram – deflected carcass, right diagram – lateral force):

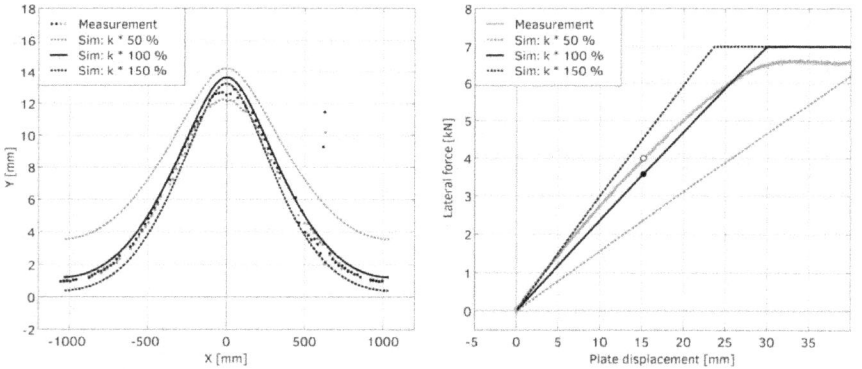

Torsional displacement maneuver (left diagram – deflected carcass, right diagram – bore torque):

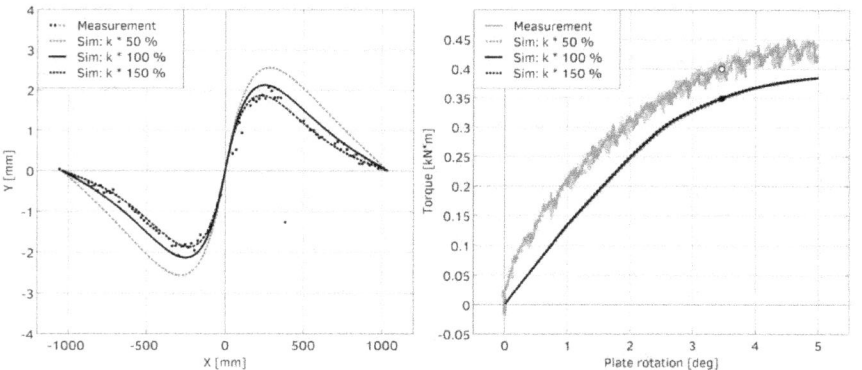

A.46. Sensitivity analysis: Width of contact patch

Slip angle step maneuver (left diagram – deflected carcass, right diagrams – force and torque):

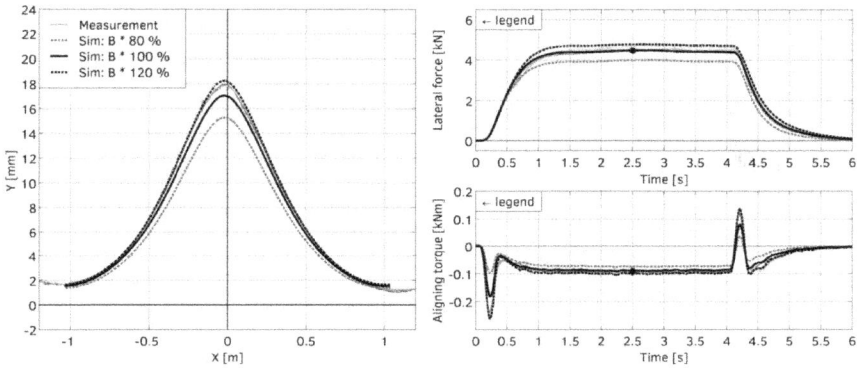

Lateral displacement maneuver (left diagram – deflected carcass, right diagram – lateral force):

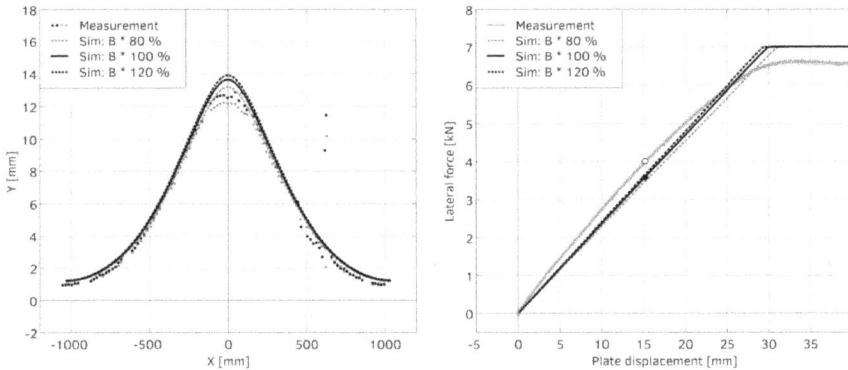

Torsional displacement maneuver (left diagram – deflected carcass, right diagram – bore torque):

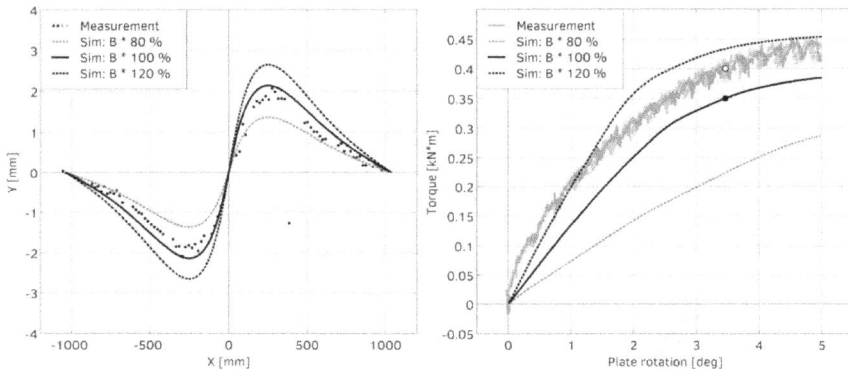

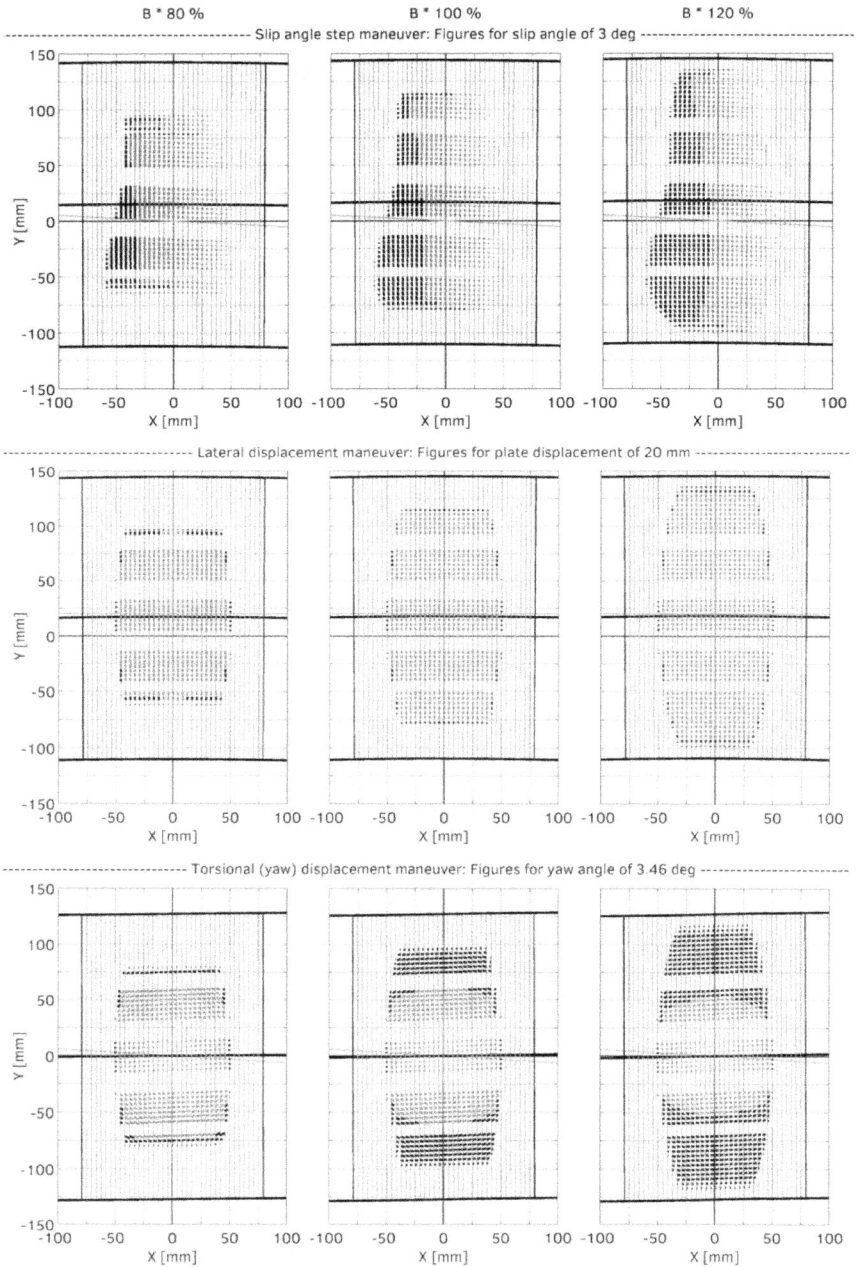

B * 80 %   B * 100 %   B * 120 %

-------------------------------- Slip angle step maneuver: Figures for slip angle of 3 deg --------------------------------

-------------------------- Lateral displacement maneuver: Figures for plate displacement of 20 mm --------------------------

-------------------------- Torsional (yaw) displacement maneuver: Figures for yaw angle of 3.46 deg --------------------------

A.47. Sensitivity analysis: Length of contact patch

Slip angle step maneuver (left diagram – deflected carcass, right diagrams – force and torque):

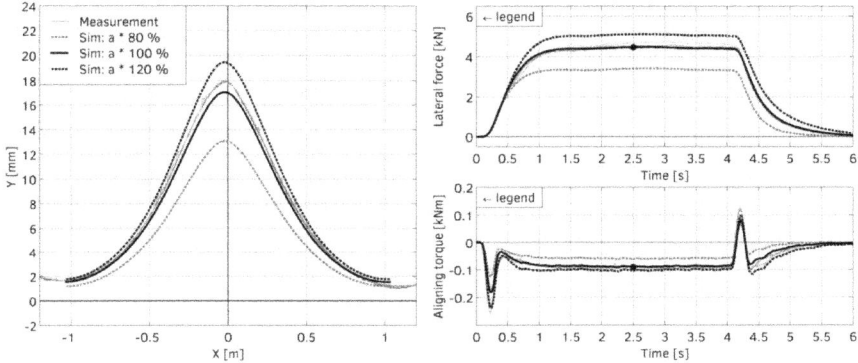

Lateral displacement maneuver (left diagram – deflected carcass, right diagram – lateral force):

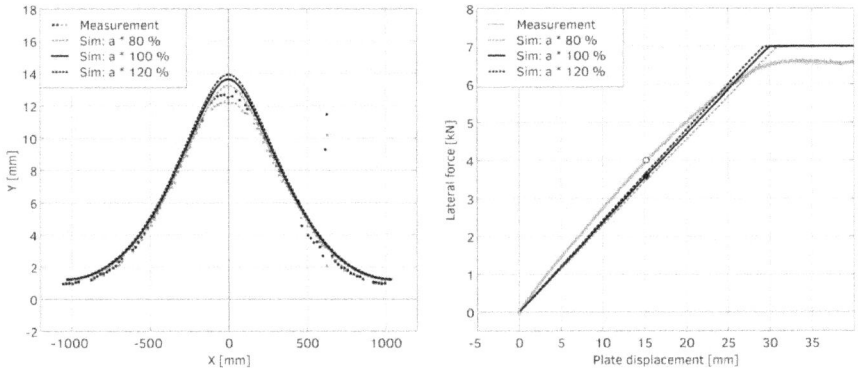

Torsional displacement maneuver (left diagram – deflected carcass, right diagram – bore torque):

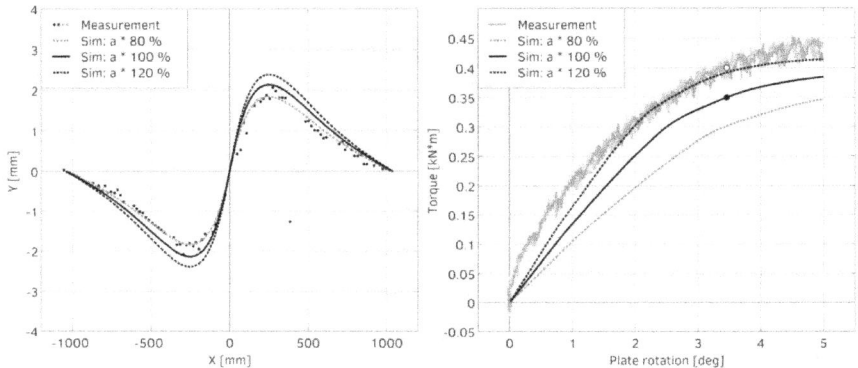

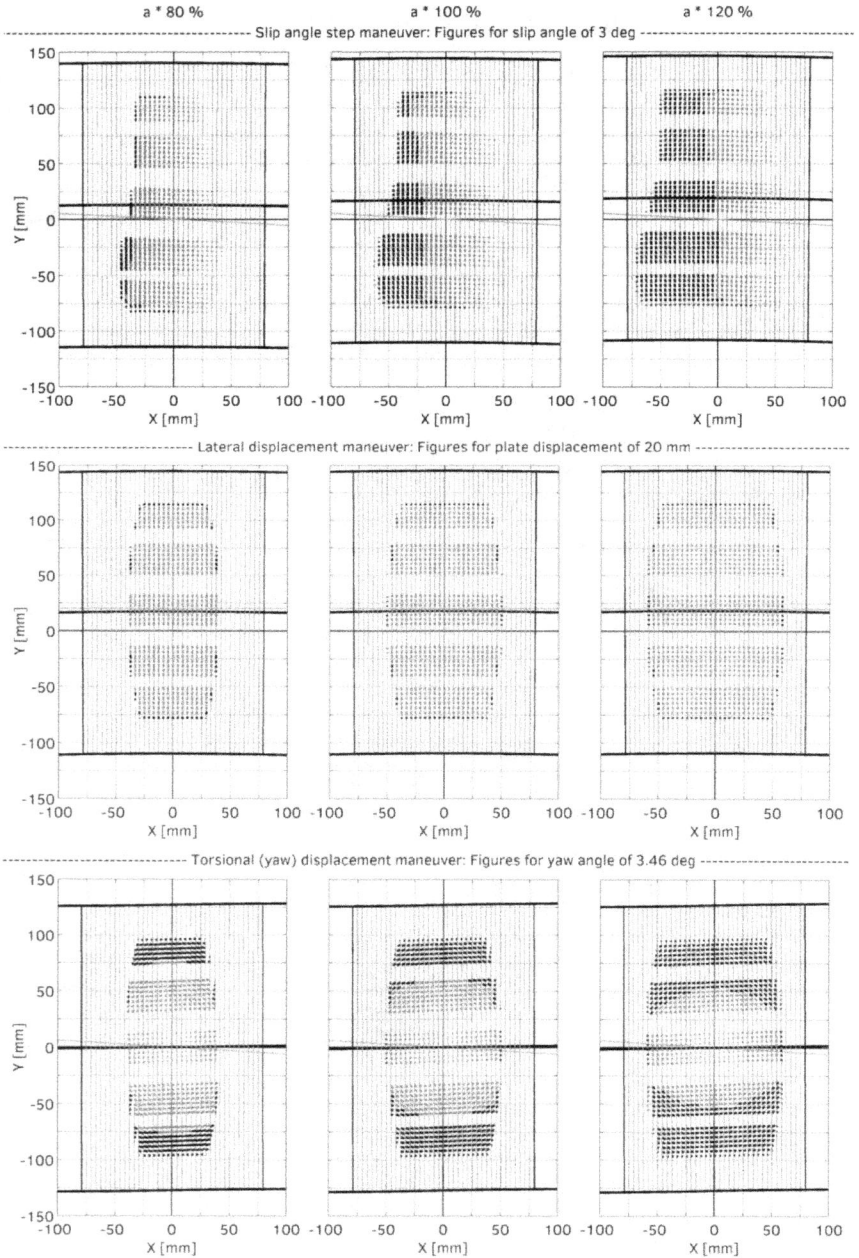

a * 80 %   a * 100 %   a * 120 %

-------- Slip angle step maneuver: Figures for slip angle of 3 deg --------

-------- Lateral displacement maneuver: Figures for plate displacement of 20 mm --------

-------- Torsional (yaw) displacement maneuver: Figures for yaw angle of 3.46 deg --------

A.48. Sensitivity analysis: Carcass shear angle coefficient

Slip angle step maneuver (left diagram – deflected carcass, right diagrams – force and torque):

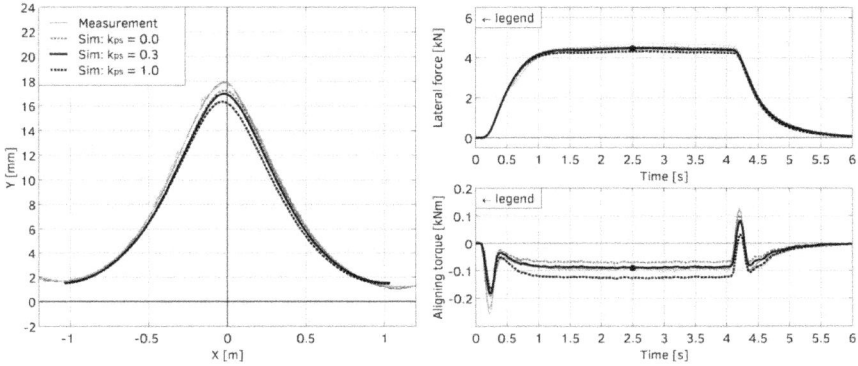

Lateral displacement maneuver (left diagram – deflected carcass, right diagram – lateral force):

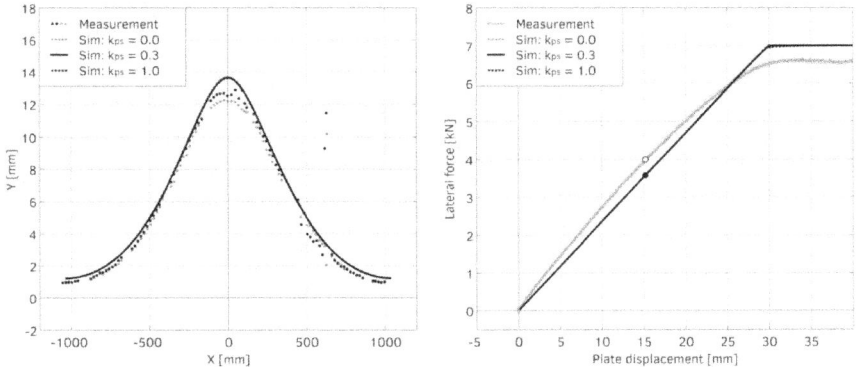

Torsional displacement maneuver (left diagram – deflected carcass, right diagram – bore torque):

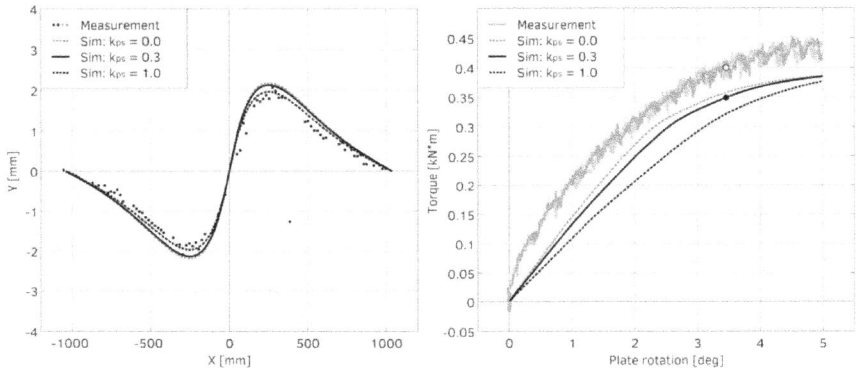

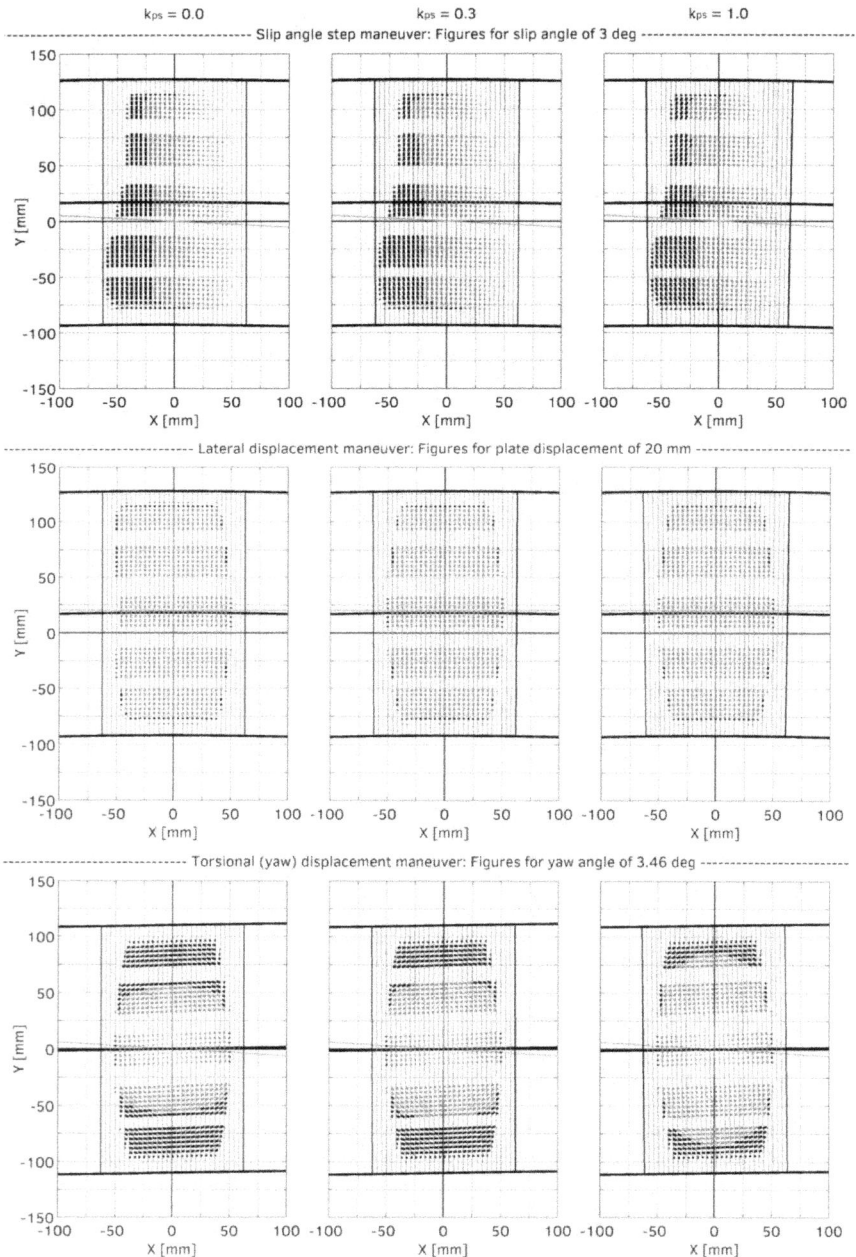

A.49. Sensitivity analysis: Variable contact patch shape

Slip angle step maneuver (left diagram – deflected carcass, right diagrams – force and torque):

A.50. Sensitivity analysis: Friction coefficient

Slip angle step maneuver (left diagram – deflected carcass, right diagrams – force and torque):

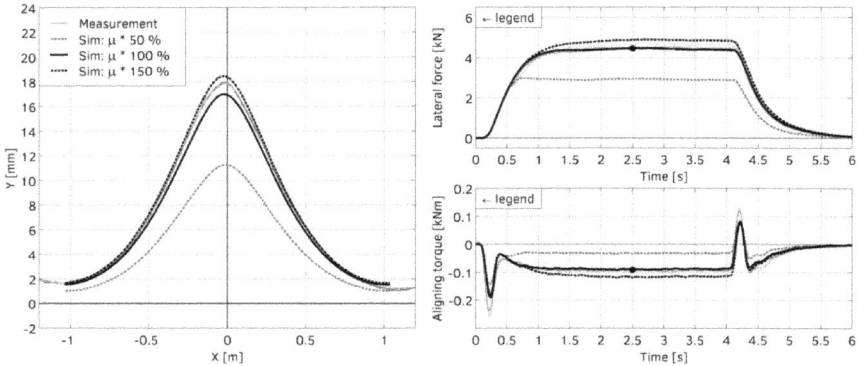

Lateral displacement maneuver (left diagram – deflected carcass, right diagram – lateral force):

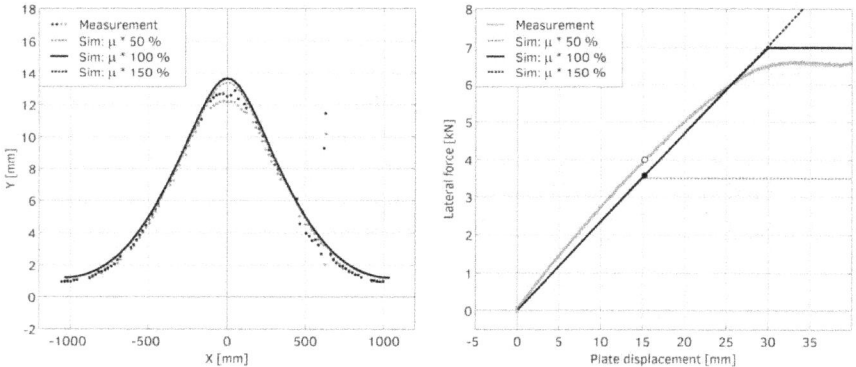

Torsional displacement maneuver (left diagram – deflected carcass, right diagram – bore torque):

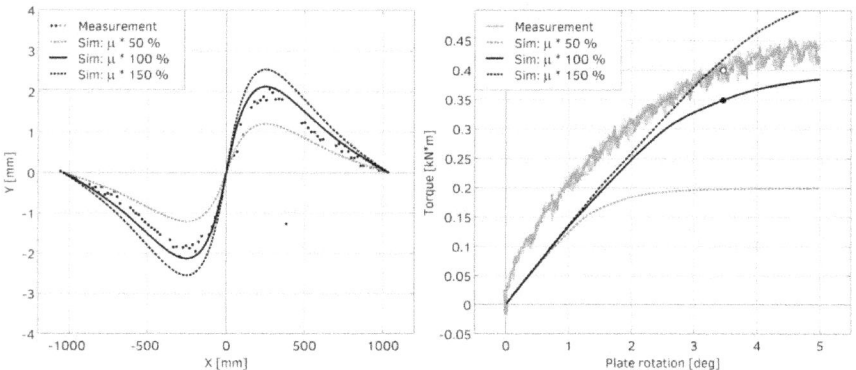

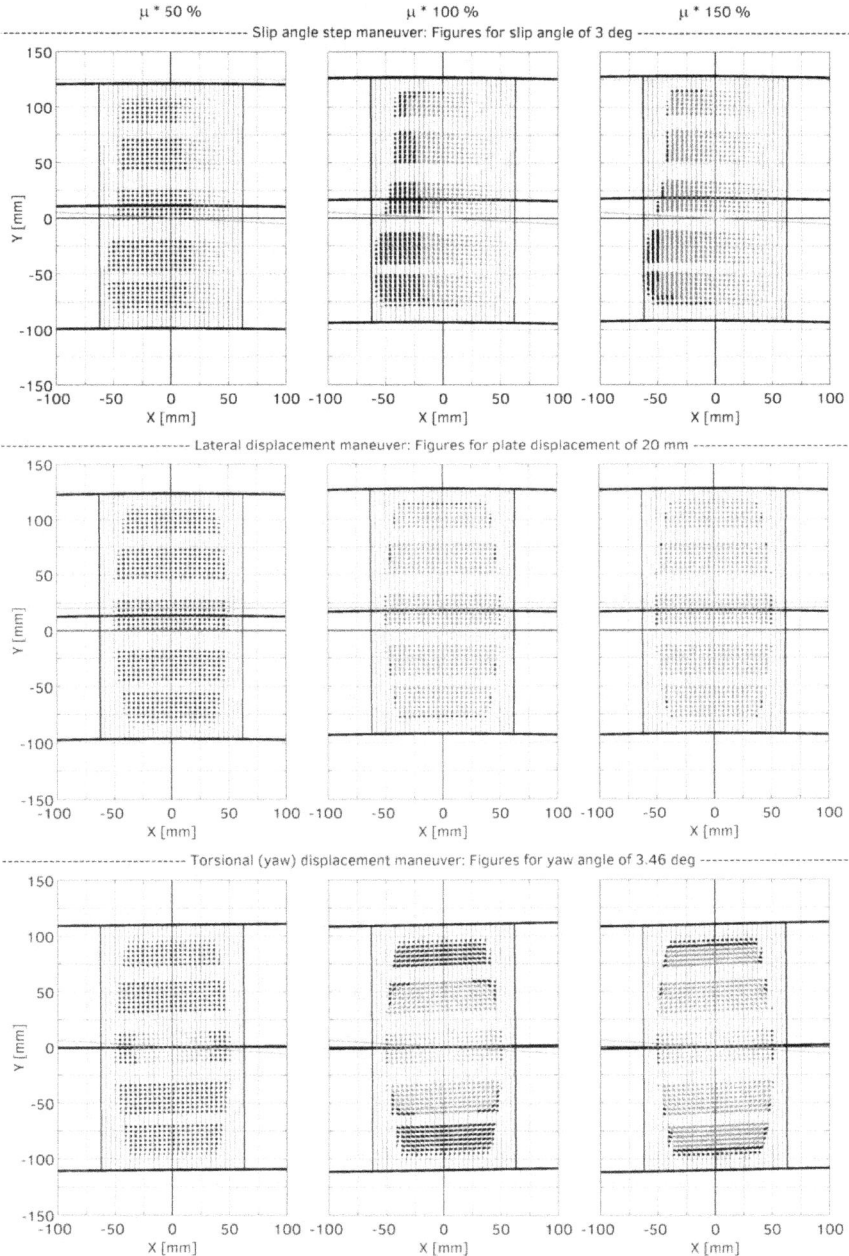

A.51. Sensitivity analysis: Grooves width in contact patch

Slip angle step maneuver (left diagram – deflected carcass, right diagrams – force and torque):

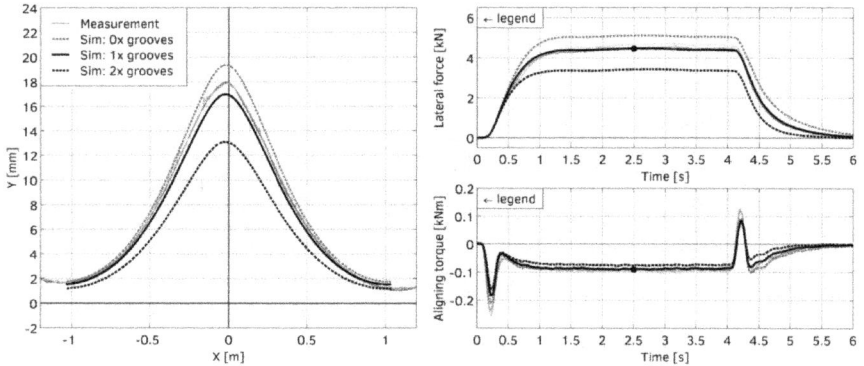

Lateral displacement maneuver (left diagram – deflected carcass, right diagram – lateral force):

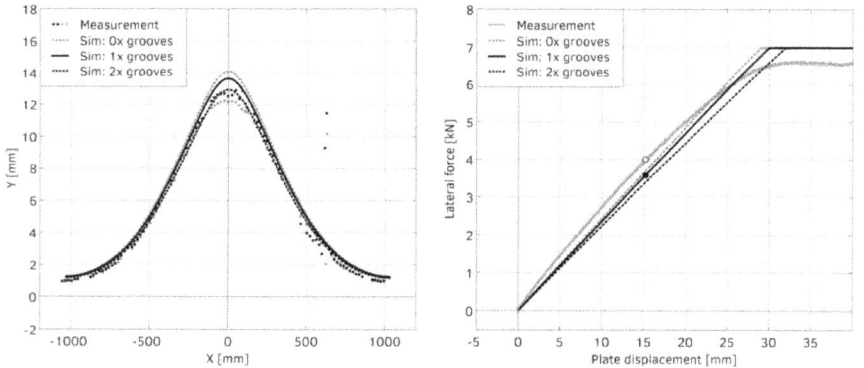

Torsional displacement maneuver (left diagram – deflected carcass, right diagram – bore torque):

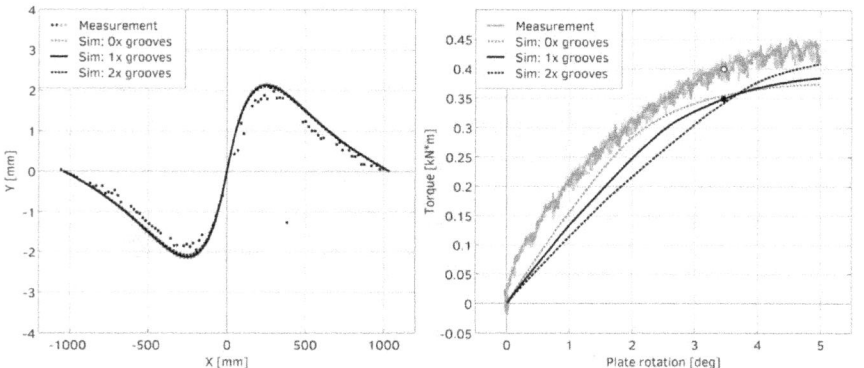

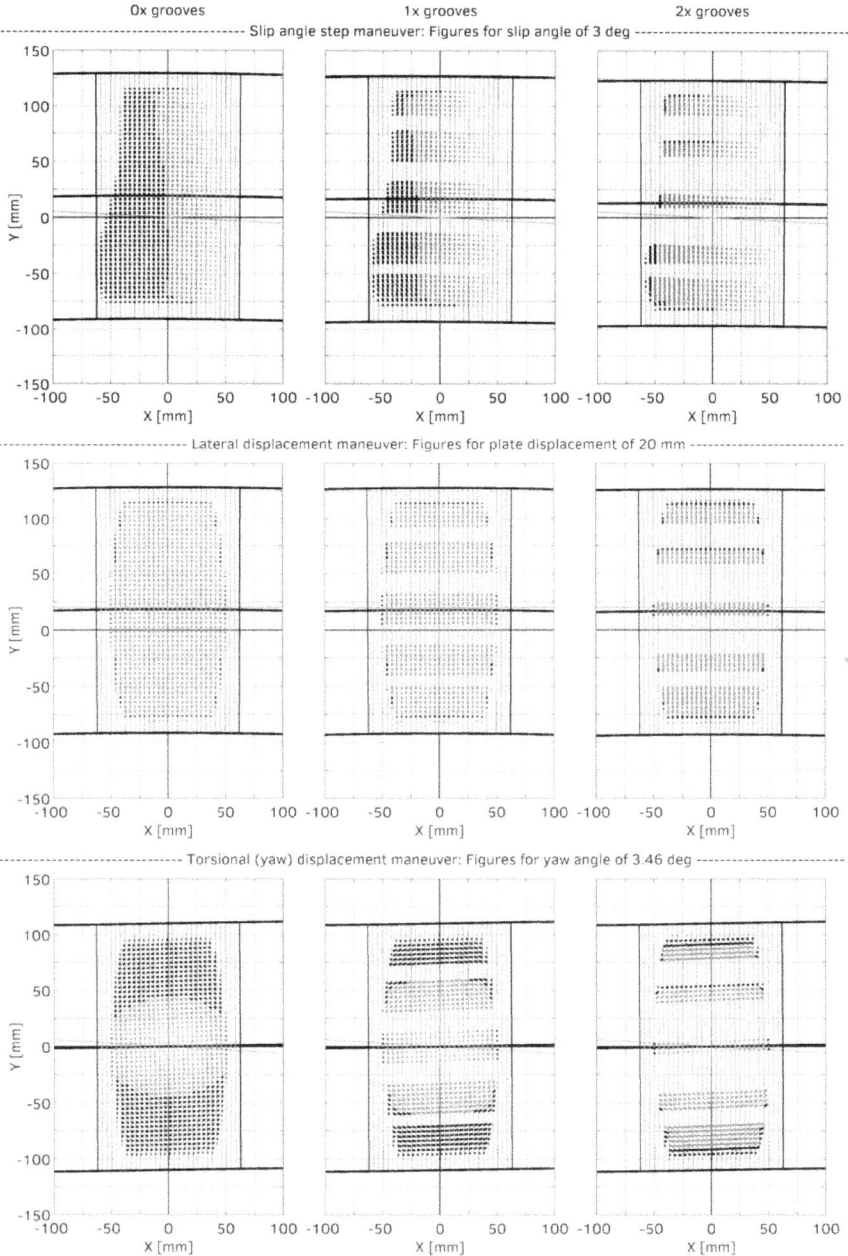

A.52. Sensitivity analysis: Distributed bending torque

Slip angle step maneuver (left diagram – deflected carcass, right diagrams – force and torque):

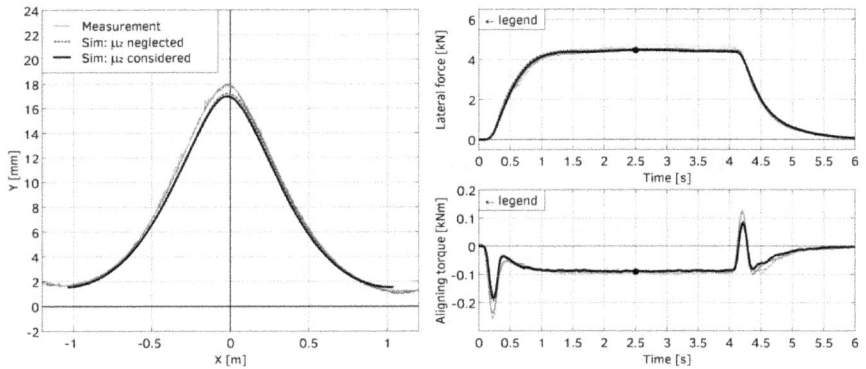

Lateral displacement maneuver (left diagram – deflected carcass, right diagram – lateral force):

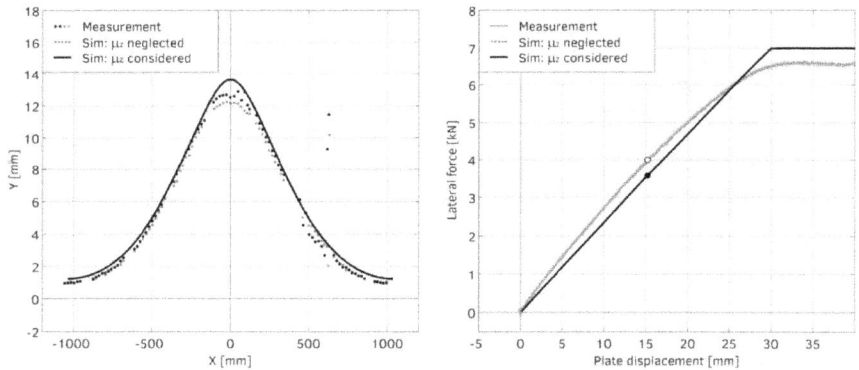

Torsional displacement maneuver (left diagram – deflected carcass, right diagram – bore torque):

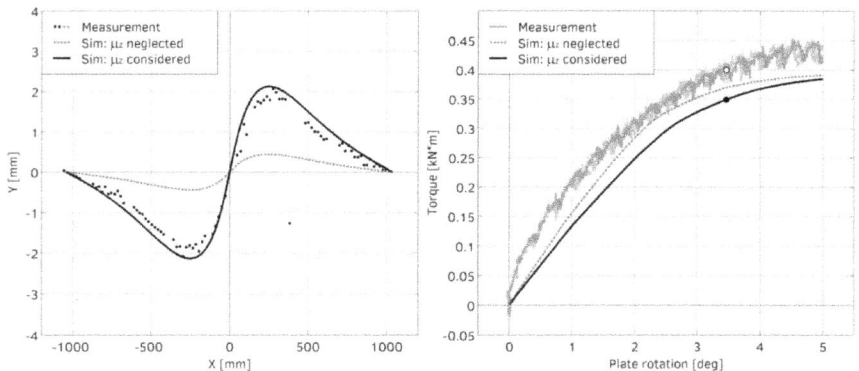

A.53. Sensitivity analysis: Number of brush elements

Slip angle step maneuver (left diagram – deflected carcass, right diagrams – force and torque):

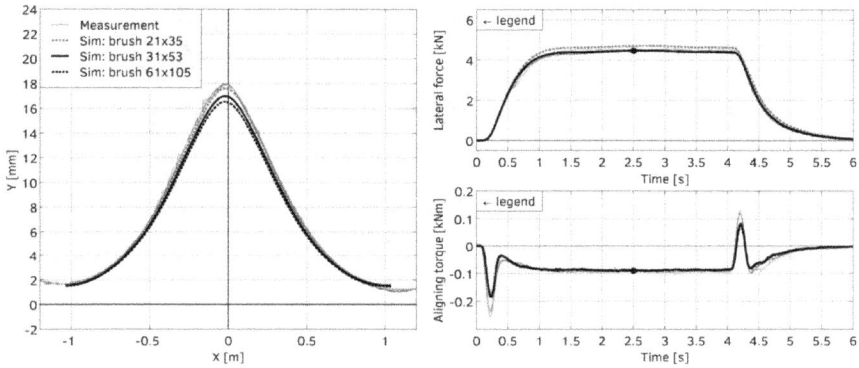

Lateral displacement maneuver (left diagram – deflected carcass, right diagram – lateral force):

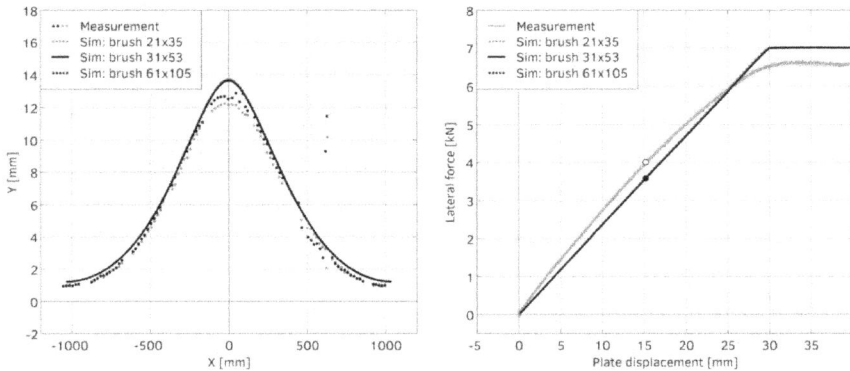

Torsional displacement maneuver (left diagram – deflected carcass, right diagram – bore torque):

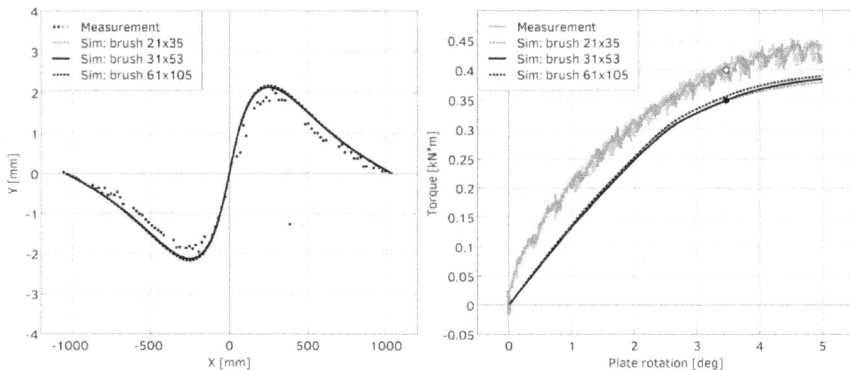

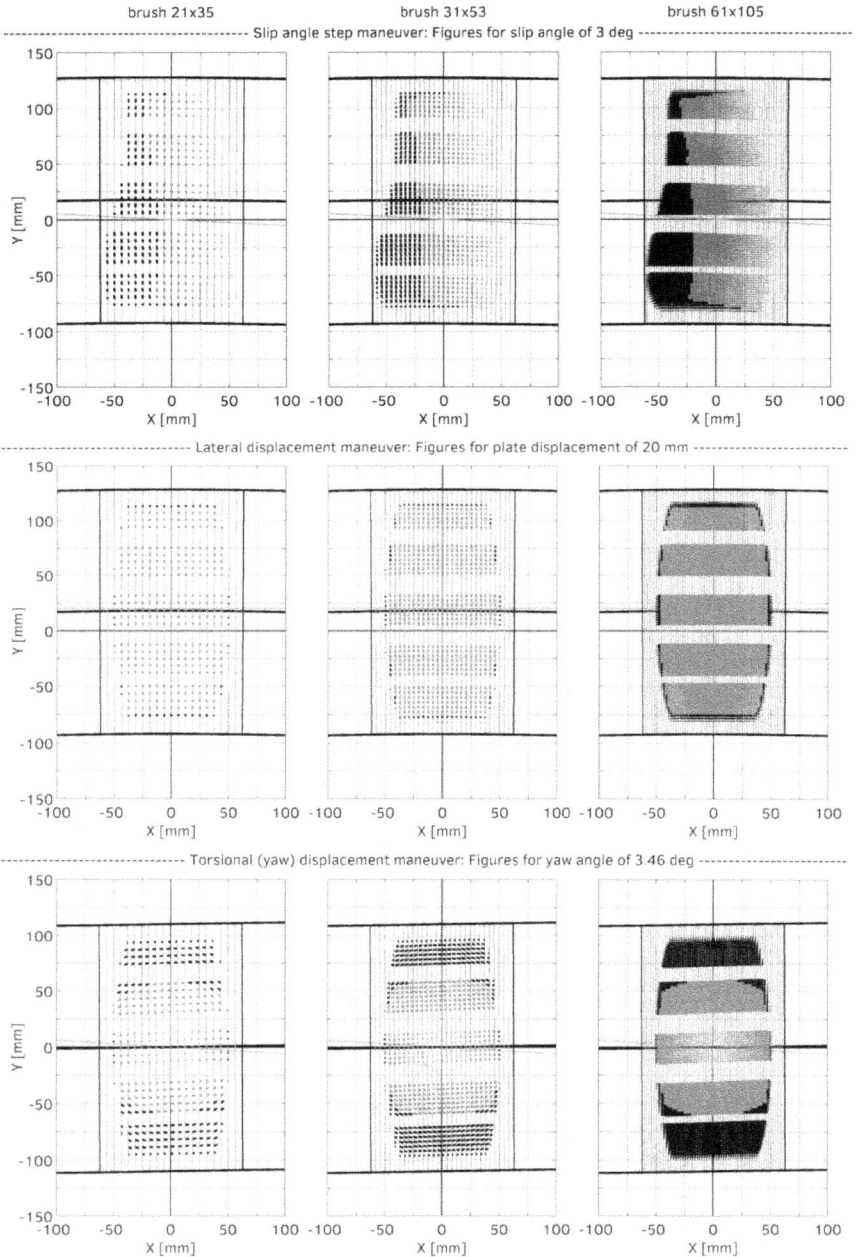

brush 21x35   brush 31x53   brush 61x105

-------------------------------- Slip angle step maneuver: Figures for slip angle of 3 deg --------------------------------

------------------------- Lateral displacement maneuver: Figures for plate displacement of 20 mm -------------------------

------------------------- Torsional (yaw) displacement maneuver: Figures for yaw angle of 3.46 deg -------------------------

A.54. Sensitivity analysis: Approximation precision

Slip angle step maneuver (left diagram – deflected carcass, right diagrams – force and torque):

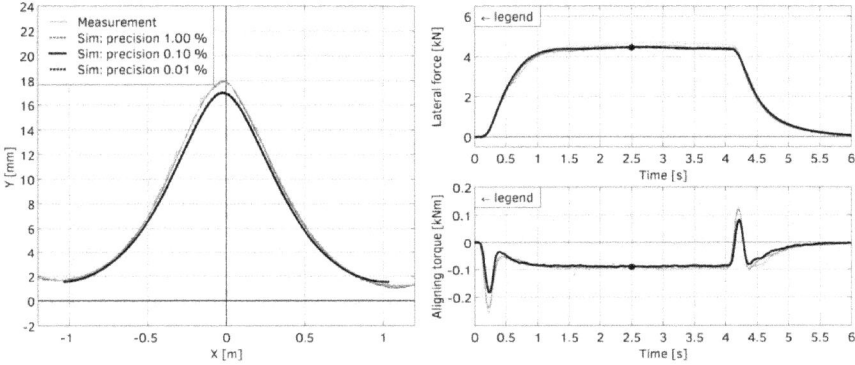

Lateral displacement maneuver (left diagram – deflected carcass, right diagram – lateral force):

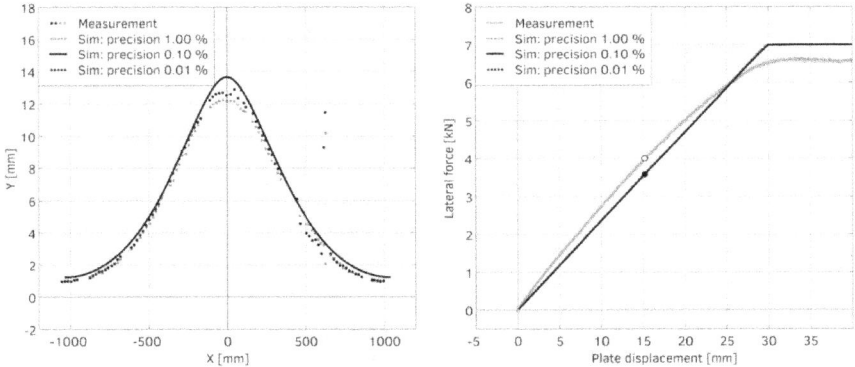

Torsional displacement maneuver (left diagram – deflected carcass, right diagram – bore torque):

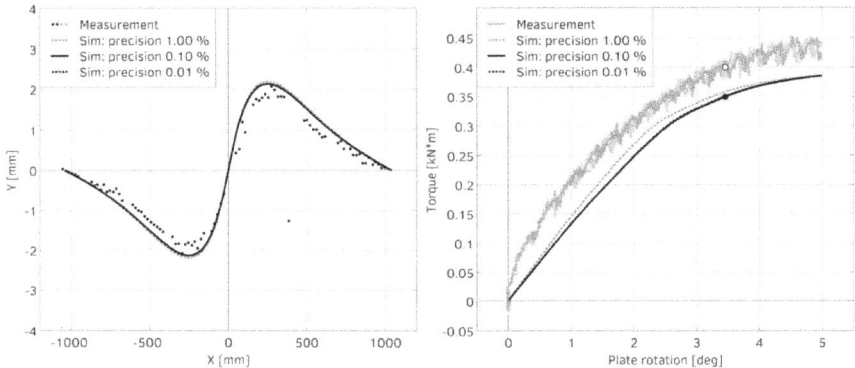

A.55. Details to the example of application

Utilized friction potential rate was analyzed based on values, obtained in the different ways. Here, these options are described in detail using the following definitions:

D   **Measured utilized friction potential rate in longitudinal (lateral) direction** k_{mx} (k_{my}) – ratio of measured longitudinal F_x (lateral F_y) force in a contact patch to the maximal friction force μF_z.

D   **Measured utilized friction potential rate** k_m – ratio of measured horizontal force $\sqrt{F_x{}^2 + F_y{}^2}$ in a contact patch to the maximal friction force μF_z.

$$k_{mx} = \frac{F_x}{\mu F_z}; \qquad k_{my} = \frac{F_y}{\mu F_z}; \qquad k_m = \sqrt{k_{mx}{}^2 + k_{my}{}^2} = \frac{\sqrt{F_x{}^2 + F_y{}^2}}{\mu F_z} \tag{A.55.1}$$

D   **Simulated utilized friction potential rate in longitudinal (lateral) direction, stress based** k_{sx} (k_{sy}) – ratio of sum of <u>magnitudes</u> of the longitudinal (lateral) shear forces of brush elements in a contact patch, calculated in simulation, to the maximal friction force μF_z.

D   **Simulated utilized friction potential rate, stress based** k_s – ratio of the sum of <u>magnitudes</u> of the shear horizontal forces of brush elements in a contact patch, calculated in simulation, to the maximal friction force μF_z.

$$k_{sx} = \frac{c_{xy} \sum_{i=1}^{N_r} \sum_{j=1}^{N_e} |d_x(i,j)|}{\mu F_z}; \qquad k_{sy} = \frac{c_{xy} \sum_{i=1}^{N_r} \sum_{j=1}^{N_e} |d_y(i,j)|}{\mu F_z}; \tag{A.55.2}$$

$$k_s = \frac{c_{xy} \sum_{i=1}^{N_r} \sum_{j=1}^{N_e} \sqrt{[d_x(i,j)]^2 + [d_y(i,j)]^2}}{\mu F_z} \tag{A.55.3}$$

D   **Simulated utilized friction potential rate in longitudinal (lateral) direction, force based** k_{fx} (k_{fy}) – ratio of the sum of <u>values</u> of the longitudinal (lateral) shear forces of brush elements in a contact patch, calculated in simulation, to the maximal friction force μF_z.

D   **Simulated utilized friction potential rate, force based** k_f – ratio of the sum of <u>values</u> of the shear horizontal forces of brush elements in a contact patch, calculated in simulation, to the maximal friction force μF_z.

$$k_{fx} = \frac{c_{xy} \sum_{i=1}^{N_r} \sum_{j=1}^{N_e} d_x(i,j)}{\mu F_z}; \qquad k_{fy} = \frac{c_{xy} \sum_{i=1}^{N_r} \sum_{j=1}^{N_e} d_y(i,j)}{\mu F_z}; \qquad k_f = \sqrt{k_{fx}{}^2 + k_{fy}{}^2} \tag{A.55.4}$$

The principle difference between stress-based and force-based rates is following: stress-based rate considers all elementary utilized forces in a contact patch (oppositely directed forces are added), whereas force-based rate considers only the sum of them (oppositely directed forces are subtracted). Hence, the stress-based rate is more precise and it is always higher than the force-based rate. However, as soon as there is no yaw turning in the contact patch, these two values are practically equal.

D   **Estimated utilized friction potential rate** k_e – a rate, calculated based on the length values of the grip l_g and slip l_s regions, assuming linear development of shear stress from the leading edge to the border between the grip and slip regions.

$$k_e = 1 - \frac{l_g}{2(l_s + l_g)}$$
(A.55.5)

The following figures depict the strain within the contact patch in different conditions as a top view of the tire carcass and contact patch (left diagram). In the top right diagram, tire excitation is shown (slip angle, brake slip). The bottom right chart illustrates the utilized friction potential rates: The simulated rate (stress based) including longitudinal and lateral components, and the estimated rate.

Based on the results of the contact patch shape change analysis, which showed practically no change in contact patch length in the middle plane of a tire depending upon camber and slip angle, it is appropriate to consider the application of only one acceleration sensor, mounted in the tire middle plane. The estimated utilized friction potential rate was calculated based on the lengths of the grip and slip regions, captured by this sensor in the middle plane of the tire.

The first observation highlighted a limitation of this method: It is unable to capture a utilized friction potential rate below 0.5, as it assumes linear shear development in a contact patch, and the perceptible slip region occurs only after the utilization of more than half of the potential.

Apart from this case, the estimation quality in a steady state was generally high, an error remained below 5 % independently, whether the longitudinal brake slip (Figure A.55.1), longitudinal drive slip (Figure A.55.2), lateral slip (Figure A.55.3) or combined slip (Figure A.55.4) is present. Hence, for the current tire and measurement conditions:

> ! Estimation of the utilized friction potential of a tire, which is based on lengths of slip and grip region, is also feasible for a cornering tire: for the lateral as well as the combined slip.

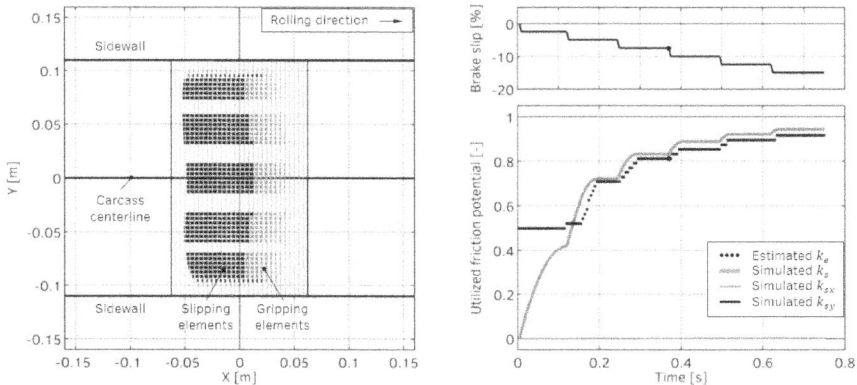

Figure A.55.1. Strain figure and utilized friction potential of the tire rolling straight ahead with brake slip excitation (strain figure is shown for a brake slip of –7.5 %).

One parameter remained to be checked to assure the implementation of the method for the cornering tire. According to this concept, the slip and grip regions were distinguished using vibration, which comes from the sensor sliding along the rough surface. If this sliding speed is not sufficiently high, then the vibration energy is low and the difference between the grip and slip region is less clear. Therefore, it was necessary to compare also the sliding speed in the slip region of the contact patch in different rolling conditions. The case of braking tire, which was proposed in [Nis17] (longitudinal slip), was selected as a reference. The cornering tire was a case of lateral slip.

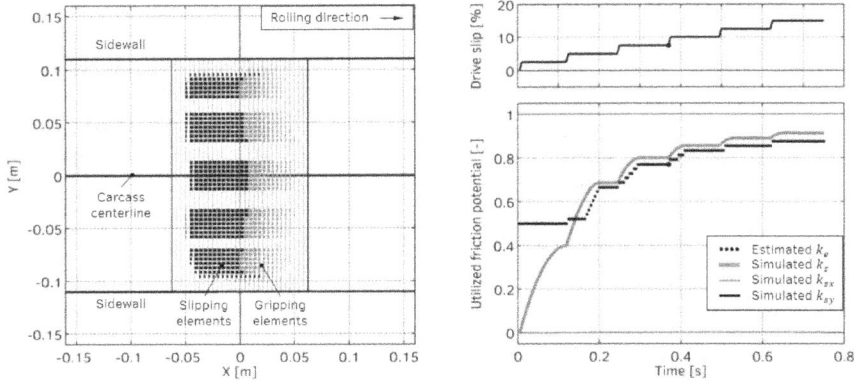

Figure A.55.2. Strain figure and utilized friction potential of the tire rolling straight ahead with drive slip excitation (strain figure is shown for a drive slip of 7.5 %).

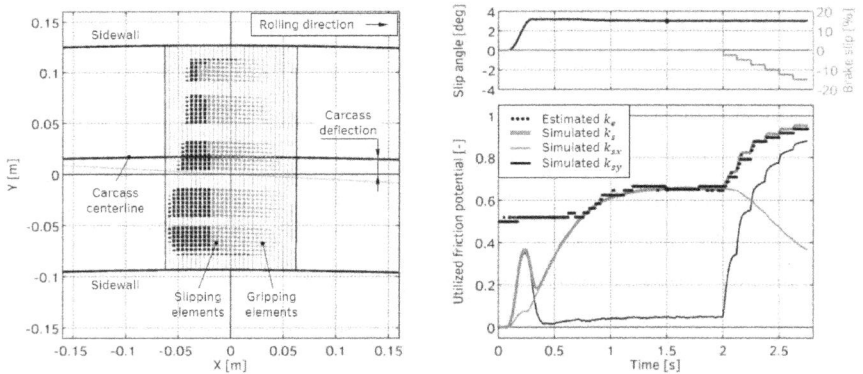

Figure A.55.3. Strain figure and utilized friction potential of the cornering tire (strain figure is shown for a slip angle of 3°, brake slip of 0 %).

In the Figure A.55.6 it is visible that sliding velocities in these two cases had the same order of magnitude. Hence, vibration frequencies had also the same order of magnitude and the sensor could distinguish the slip and grip regions also in the cornering tire.

It was determined that the sliding speed in the sliding region of the contact patch of a tire, cornering with a slip angle of 3°, corresponded to the sliding speed of the straight ahead rolling tire with a brake slip of −5 %. As far as both values belonged to the linear range of longitudinal or lateral force generation, it is appropriate to confirm the reasonability of the application of the described estimation method not only for braking, but also for cornering tire.

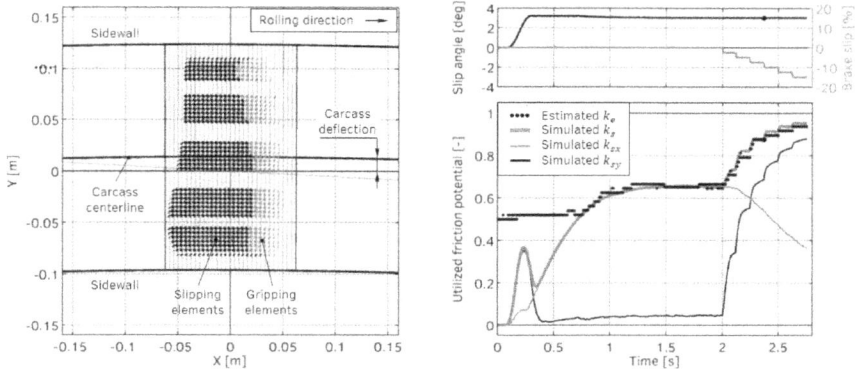

Figure A.55.4. Strain figure and utilized friction potential of the cornering tire excited with brake slip (strain figure is shown for a slip angle of 3°, brake slip of –7.5 %).

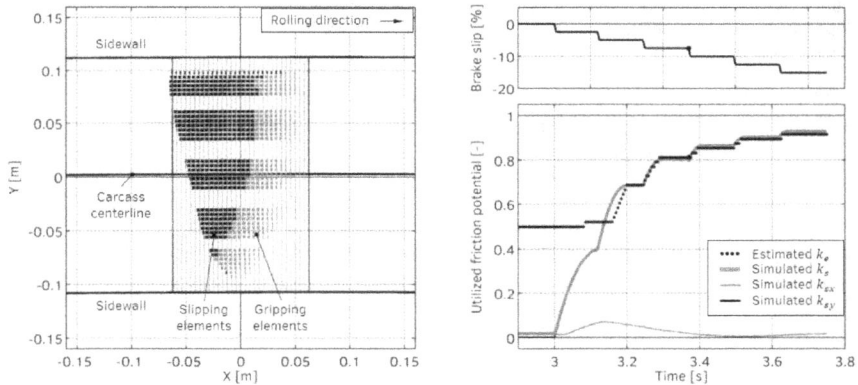

Figure A.55.5. Strain figure and utilized friction potential of the cambered tire excited with brake slip (strain figure is shown for a camber angle of –4°, brake slip of –7.5 %).

The important insights, generated based on the model for the given tire and rolling conditions, can be summarized in the following way:

! Although the strain figures featured asymmetry, one sensor located in the middle plane of the given tire provided an estimation of the utilized friction potential rate in a steady state in the range of 0.5-1.0 with an error below 5 % for longitudinal slip (braking tire), lateral slip (cornering tire) and combined slip.

! The speed of tread sliding along the road surface in the sliding region of the contact patch of the cornering tire with a slip angle of 3° corresponded to the sliding speed of the straight ahead rolling tire with a brake slip of –5 %.

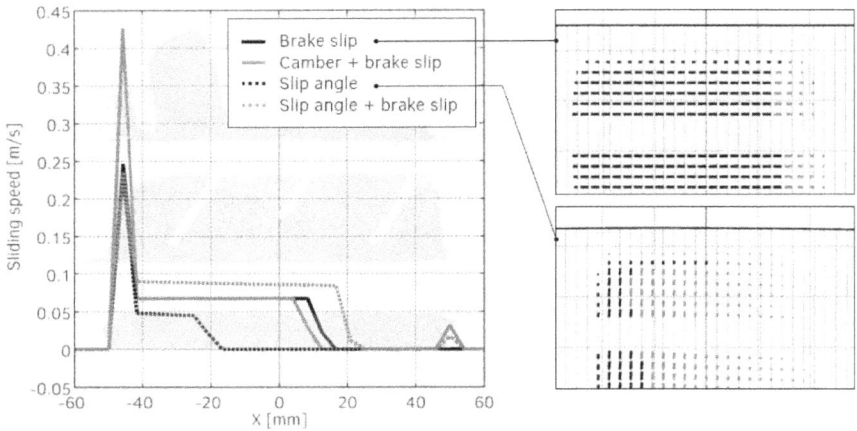

Figure A.55.6. A comparison of sliding speed values in the sliding region of a contact patch depending upon the rolling conditions (for different curves the brake slip is −7.5 %, slip angle is 3°, camber accounts for −4°).

Hence, application of the described estimation method is feasible not only for braking, but also for cornering tire. However, this result should not be generalized for all tires. It is used in the thesis only to illustrate the possible application of the developed knowledge.

According to the trends of automotive industry, the next task is to expand this method in order to be able to estimate, how high is longitudinal and how high is lateral component of the utilized friction potential rate. This provides an insight, which intervention on the tire increases its force response without approaching friction limit: steering intervention (lateral slip) or brake intervention (longitudinal slip). Such a functionality can be helpful for advanced driver assistance systems. Hence, this challenge serves as a promising objective for further research.

www.ingramcontent.com/pod-product-compliance
Lightning Source LLC
Chambersburg PA
CBHW060308220326
41598CB00027B/4273